Much Ado about (Practically) Nothing

Books by David E. Fisher

Novels

Crisis
Compartments
A Fearful Symmetry
The Last Flying Tiger
The Man You Sleep With
Variation on a Theme
Katie's Terror
Hostage One
The Man

Nonfiction

The Creation of the Universe
The Creation of Atoms and Stars
The Ideas of Einstein (Juvenile)
The Third Experiment
The Birth of the Earth
A Dance on the Edge of Time: The Birth of Radar
The Origin and Evolution of Our Own Particular Universe
Fire and Ice: The Greenhouse Effect, Ozone Depletion, and Nuclear Winter
Across the Top of the World
The Scariest Place in the World
Tube: The Invention of Television (with M. J. Fisher)
Strangers in the Night: A Brief History of Life on Other Worlds (with M. J. Fisher)
Mysteries of the Past: Companion to the NOVA series (with M. J. Fisher)
A Summer Bright and Terrible

DAVID E. FISHER

Much Ado about (Practically) Nothing

A History of the Noble Gases

OXFORD
UNIVERSITY PRESS
2010

OXFORD
UNIVERSITY PRESS

Oxford University Press, Inc., publishes works that further
Oxford University's objective of excellence
in research, scholarship, and education.

Oxford New York
Auckland Cape Town Dar es Salaam Hong Kong Karachi
Kuala Lumpur Madrid Melbourne Mexico City Nairobi
New Delhi Shanghai Taipei Toronto

With offices in
Argentina Austria Brazil Chile Czech Republic France Greece
Guatemala Hungary Italy Japan Poland Portugal Singapore
South Korea Switzerland Thailand Turkey Ukraine Vietnam

Published by Oxford University Press, Inc.
198 Madison Avenue, New York, NY 10016

www.oup.com

Library of Congress Cataloging-in-Publication Data
Fisher, David E., 1932–
Much ado about (practically) nothing : a history of the noble gases / David E. Fisher.
p. cm.
Includes bibliographical references and index.
ISBN 978-0-19-539396-5
1. Gases, Rare. I. Title.
QD162.F57 2010
546'.75—dc22 2009054365

9 8 7 6 5 4 3 2 1

Printed in the United States of America
on acid-free paper

For
Jackson Henry
and
Louis Samuel

What is the path? There is no path. On into the Unknown!

—Niels Bohr, loosely translated from Goethe's *Faust*

Acknowledgments

You do this, and you do that, and suddenly that's your life.
—Bernhard Schlink

My gratitude is due to Alex Zucker and John Pinajian, who taught me science, to Ollie Schaeffer and Ray Davis, who introduced me to the noble gases, and to Tommy Gold and Cesare Emiliani for expanding my horizons. I am indebted to Neta Bahcall, Grenville Turner, Robert Fleischer, and Norbert Porile for their suggestions regarding this book, although they must be held blameless for the finished product; it should be clear that all opinions expressed herein are mine alone. Conversations reported with these and others over the past fifty years are accurate to the best of my recollection, but just as space is warped by mass, memory is warped by time; some slack must be cut.

Contents

Much Ado about (Practically) Nothing

ONE

Philosophy and Apology

The most incomprehensible thing about the universe is that it is comprehensible.

—Albert Einstein

TO PARAPHRASE THE FIRST ADVERTISEMENT FOR MY first book ("*Crisis* is a terrifying novel. But don't let that scare you"), this book is about the noble gases. But don't let that scare you. It's really about how science works. There is a general misapprehension about this. Most people, without thinking about it, visualize the universe as a railroad track disappearing into the distance, and science as the locomotive slowing wending its way along the track, learning year by year more and more about this universe in which we live.

Not so.

It would be more realistic to visualize the universe as a black forest hidden on a cloud-obscured night, with science as a lost child trying to find its way home, feeling blindly the branches of the trees, occasionally being slapped in the face by one, tripping over the roots of another, stumbling on a path and taking it eagerly only to find it branching or, worse, precipitately ending. Nothing to do then but turn around and go back, find another branch, another path, or, worse luck, with no path to be found, try again and again to feel your way through the dark trees striving to find some light, somewhere, anywhere.

The only thing wrong with this analogy is that being lost in such a forest would be terrifying, whereas science is fun. What is right about the analogy is that science does not run along a straight path like the locomotive but bumbles to the right and left, sometimes backwards, and every once in a while takes a step closer to home, to the ultimate goal, to an understanding of our universe.[1]

The last part of that sentence, if you think about it, is astounding. Despite being born naked and ignorant of everything around us we have learned from solely our own efforts that this flat ground we walk on is actually curved, part of a spheroid, that the stars we see are suns, that everything we touch and hold is made up of a hundred or so different particles, that our world has existed not forever but for four and a half billion years, and that many of the stars are billions of years older, in fact that the entire universe is just under fourteen billion years old. This and so much more we know; a truly amazing feat, expressed best by the quote which opens this section—but another quote (by J. B. S. Haldane) serves to balance it: "The Universe is not only queerer than we suppose, but queerer than we *can* suppose."[2]

Somewhere between these two statements we stir and stumble around, trying to make sense of it; this is what I'm trying to write about, and a most excellent illustration of it all is the story of the discovery and uses of the noble gases, a group of elements vanishingly rare on our planet, which do not react with anything, which in fact do nothing. You can't feel them, hear them, see them or smell them. Thus their names: the rare gases, for their rarity; the inert gases, for their inability to form compounds; finally, the noble gases, for their ability—like the nobility—to exist without doing any work.

But for all this, starting with the discovery of the rarest of these invisible and seemingly useless gases, they have turned out to be

1. I say "our" universe because we don't even know if there are others.

2. A slightly more pessimistic view was expressed by Woody Allen: "Nothing worth knowing can be understood."

instrumental in understanding our universe, from determining its age to learning what makes the stars shine and perhaps to the origin of life itself.

Thus the title of this book: much ado has been both caused by and focused on a group of somethings which are practically nothing—and which illustrate the ebb and flow (and the occasional tsunami) of the tide of science.

Finally, an apology. It is not possible to write a comprehensive, in-depth review of all the uses to which the rare gases have been put. For example, if you want to read about recent advances in the field of geo/cosmochemistry alone, without the historical stories, there is an 844-page book available, plus several chapters in another. If instead you just want to read about the discovery and uses of liquid helium, there's a fine book of just under 650 pages, plus another (500 pages, packed with equations) with the intriguing title *The Universe in a Helium Droplet*, which uses the properties of supercooled helium to "give an insight into trans-Planckian physics and thus helps in solving the cosmological constant problem and other outstanding problems in high-energy physics and cosmology." Or you might want to spend 135 dollars for a book on just two of the noble gas isotopes. Other books abound on the astronomical relations, the environment, the age determinations, the past and the future. This book is an attempt to portray the most important aspects of the story in a readable (i.e., jargonless) manner, along with an account of my fifty years with the gases and people met along the way. Suitable references for the topics discussed (and the books mentioned above) are given in the Notes section for those happy few who wish to delve more deeply.

TWO

In the Beginning

The answer, my friend,
Is blowing in the wind...

—Bob Dylan

I WAS FORMALLY INTRODUCED TO THE NOBLE GASES in high school chemistry class, but they were a boring lot and we mostly ignored them. I don't remember them even being mentioned during college or graduate school. My real affair began in the fall of 1958 at the Brookhaven National Laboratory.

I had taken a postdoctoral appointment there to work with one of the foremost nuclear chemists in the country. In 1958 there was Glenn Seaborg in California, Nate Sugarman at the University of Chicago, and Gert Friedlander at Brookhaven. I had finished my PhD work on nuclear reactions at Oak Ridge, and applied to both Chicago and Brookhaven—California was just too far away—with Brookhaven as my clear favorite: it had the world's biggest atom smasher, the Cosmotron, with 3 Bev of energy, while Chicago had only a few-hundred-Mev machine.[1]

So in the spring of 1958 I set off on my first interview trip, hitting Brookhaven first. One of the staff scientists took me around the lab, introducing me to the others. Whenever someone tried

1. MeV is a million electron volts, a unit of energy. BeV is a billion eV (now called GeV, giga eV, to avoid confusion, because the word *billion* in England means a thousand times what the rest of the world calls a billion).

to ask me about my research, he shut them off, which I thought strange. And then at eleven o'clock we walked through a door and suddenly there was a room filled with the entire chemistry department looking at me, and my host stopped at the lectern just long enough to say, "Our speaker today is David Fisher, who will tell us about his work on the nuclear reactions induced by nitrogen on sulfur." Everyone clapped politely, my host sat down, and I stood there like an idiot.

I hadn't had any idea I'd be expected to give a formal lecture to the greatest scientists I had ever met. I hadn't prepared anything. I quickly considered my options. I could turn and flee through the door I had entered. I could drop dead. I could—

There wasn't anything else. Those were my only options.

Or, oh yeah, I could begin to talk.

Actually it went very well. I had finished my research by this time—I hadn't realized how lucky I had been at Oak Ridge; everything had gone well, which is not the usual way of research—and I knew what I was talking about. So I gave a pretty good talk until the first question, which was "How do you quantify the nitrogen flux?"

I didn't know. At Oak Ridge I would prepare a sulfur target and give it to the cyclotron crew and they'd radiate it with 27-MeV nitrogen ions and then tell me the flux. I had no idea this was a difficult thing to do, that the Cosmotron people were trying various methods, and that the Oak Ridge people had worked out a good way. I did remember one thing that had happened my first week at Oak Ridge. I was wandering around the lab, orienting myself, opening doors and peering in, just looking around. And then on my first Friday I was sitting at my desk when Alex Zucker, the head of the program, opened one of the doors, walked in—and came roaring out a moment later, shouting "Who did that?"

It was like Zeus tearing down Olympus and calling out the thunder of the heavens. I didn't know what was happening.

"Who opened the door?!" Zucker roared.

Oh God. I had opened that door a little while before, looked into the darkened room, turned on the light, saw nothing interesting, and closed the door.

I thought for one brief moment of saying nothing, but then I stood up. "I did," I croaked.

Dr. Zucker looked at me, thunder and lightning swirling around him, apoplexy rising into his eyes . . . and then without a word he turned around and strode out of the room. The others explained: They ran the cyclotron all week, and had devised a photographic method of measuring the flux. At the end of the week, Friday, they took out the week's film and developed it—in the dark room I had earlier visited. They used to have a "Keep Out!" sign they put on the door while the film was developing, but they had stopped using it because everyone knew to be careful on Friday.

Everyone but me. I had ruined the film, and made the whole lab's work of the entire week useless. So while Zucker was looking at me, while the thunder and lightning were booming, two thoughts were bouncing against each other in his head: Fisher had ruined everything, and no one had told Fisher about the dark room . . . I don't think I could have been as forgiving as he was.

So now, at Brookhaven, I told them it was a photographic technique, but I didn't know anything about it.

At Chicago things went much better. On the plane from New York I went over what I had done at Brookhaven, organized it better, and at the now-expected seminar no one asked me about the flux. So everything was fine—until the seminar was over. It had been in the evening and now, about nine o'clock, Sugarman and the other faculty said they'd walk me back to my hotel. I said it wasn't necessary, but they insisted. As we walked down the trash-littered streets around the university they explained: "You don't walk alone at night here."

Flying home I figured I had a good chance at Chicago, none at Brookhaven. But how could I bring my wife to live in a place where the trash blew in the wind and you couldn't walk alone at night?

The next week I got a letter from Sugarman. He had asked me what my future plans were, and I told him I loved research and wanted to concentrate on it, probably at a national lab where I could do it full-time. He wrote that he considered his post-doc position to be a preparation for university life, and so he was offering it to another candidate, who was planning an academic career, instead of to me. That same day I was offered the position at Brookhaven.[2]

I got to Brookhaven in June, to find that Friedlander had just left for a year's sabbatical in England. "Look around, find someone else to work with," I was told. Instead, I spent the summer at nearby Westhampton Beach, surfing and sunning and sulking, feeling badly used by his desertion. When the weather chilled, so did I; I wandered around the department, and found the noble gases.

Well, what I found was a jolly little man named Ollie Schaeffer who had a home-built mass spectrometer. It was a clever contraption: the idea was that you would put a sample into a glass-enclosed vacuum system and pump out all the air, then melt or vaporize the sample with radio frequency induction heating, similar to but much more powerful than today's microwave ovens. This released the gases, which were then pumped over various chemicals to react and form nongaseous compounds—everything except the noble gases, which don't react with anything. They were then pumped into the mass spec and ionized by a jolt of high-energy electrons, and then this beam of ions was shot down the mass spec tube by a blast of voltage. On their journey, their path was deflected by a variable magnet and steered into a collector.

By varying the electron ionizing energy, the kick voltage, and primarily the magnet strength, he was able to separate the noble

2. The other guy ended up at Dupont when his fellowship was over; I spent my entire subsequent career at universities. Go figure.

gas ions according to their masses, and steer them one by one into the collector. So I was thinking...

At Oak Ridge I had inserted a target into the cyclotron, bombarded it and split the atoms apart into a bunch of isotopes, then chemically separated a single isotope and measured its production rate. By measuring the relative production rates of several of these as the target atom was smashed apart, I could investigate its nuclear structure, sort of like a little kid smashing his father's watch with a hammer to see what's inside. It took a week's work to measure one isotope. But now I could put a target into the Cosmotron, and after the bombardment I could measure all the isotopes of all the noble gases in one day: two of helium, three apiece of neon and argon, and a whole goddamn bunch of krypton and xenon isotopes. Wowee! I had visions of completing a year's research in one day...

Yeah, right. It didn't exactly turn out that way.

THREE

Helium

"He isn't backward. He's a scientist."
"Oh, is that what it was? I knew he was . . . peculiar."
—*The Lady Eve*, by Preston Sturges

TODAY WE LEARN AT SUCH A YOUNG AGE ABOUT the periodic properties of the elements and their atomic structure that it seems as if we grew up with the knowledge, and that everyone must always have known such basic, simple stuff. But till nearly the end of the nineteenth century no one even suspected that such things as the noble gases, with their filled electronic orbits, might exist.

Helium was the first one we at Brookhaven looked for in our mass spectrometer, and the first one discovered. This was in 1868, but the discovery was ignored and the discoverer ridiculed. He didn't care; he had other things on his mind.

His name was Pierre Jules César Janssen, and he was a French astronomer who sailed to India that year in order to take advantage of a predicted solar eclipse. With the overwhelming brightness of the sun's disk blocked by the moon, he hoped to observe the outer layers using the newly discovered technique of absorption spectroscopy. Nobody at the time understood why, but it had been observed that when a bright light shone through a gas, the chemical elements in the gas absorbed the light at specific wavelengths. The resulting dark lines in the emission spectrum of the light were like fingerprints, for it had been found in chemical laboratories that when an element was heated it emitted light at the same wavelengths it would absorb when light from an outside source was shined on it. So the way the technique worked, Janssen reasoned, was that he could measure the wavelengths of the solar absorbed lines and compare them with lines emitted in chemical laboratories where different elements were routinely studied, thus identifying the gases present in the sun. On August 18 of that year the moon moved properly into position, and Janssen's spectroscope captured the dark absorption lines of the gases surrounding the sun.

It was an exciting moment, as for the first time the old riddle could be answered: "Twinkle twinkle, little star, how I wonder what you are." The answer now was clear: the sun, a typical star, was made overwhelmingly of hydrogen. But to Janssen's surprise there was one additional and annoying line, with a wavelength of 587.49 nanometers. It was annoying because, unlike the other lines he found, it had no corresponding line among the known elements. After giving it some thought his annoyance morphed into triumph and he announced that he had found a new element in the sun, an idea so ridiculous that it evoked laughter in labs around the world.

Astronomers were used to such inanities by this time. A hundred years earlier William Herschel, the discoverer of Uranus, had seen trees growing on the moon. Stories of Martian and Venusian cities were commonplace, as were all kinds of alien hoaxes. A new element in the sun seemed just another one of these fairy tales, and so they laughed.

Janssen didn't worry about their reaction; he was much too busy. He had already traveled to Peru to study earth's equator and measured solar rays in Italy and Switzerland. He had worked on magnetic phenomena in the Azores and seen the transits of Venus in both Japan and Algeria. He had studied other eclipses from Spain to Siam, and by the time he was being laughed at for his effrontery in finding a new element in the sun he was back in a Paris surrounded by the armies of Prussia. It was now

1870—scientific news took a while to travel in those days—and the siege of Paris was in full swing. It was impossible to get out of the city, but he had business elsewhere: another eclipse was imminent in northern Siberia. So he took to the air.

The air was already being used in warfare, albeit sporadically. Some twenty years earlier the Austrian army, attempting to breach the barricades erected around Venice, loaded bombs onto a few hundred hot-air balloons and sent them aloft to drift over the beleaguered city. They had equipped the balloons with primitive fuses—actually, the height of advanced technology for the time—which were supposed to release the bombs when the wind brought the balloons over the city limits. Unfortunately the Austrian calculations were upset by a recalcitrant wind which reversed direction and brought the balloons back over their own troops just as the fuses released the bombs. Well, back to the old drawing board.

But not for the Austrians, who gave up on the idea of aerial warfare. The French, however, who had been the first people in the world to travel through the air, took up the challenge. By 1870 they had an organized troop of balloonists, called aeronauts, and when Paris became completely surrounded by the Prussians, they suggested to the head of the Post Office that balloons should be used to communicate with the outside world. The Post Office accepted the suggestion, and on September 23 the professional aeronaut Jules Durouf departed from the Place Saint-Pierre in Montmartre in the balloon Le Neptune with 227 pounds of mail. He landed his balloon safely three hours and fifteen minutes later behind enemy lines at the Chateau de Craconville. On his way, Durouf dropped visiting cards on the enemy position as he flew above the reach of their guns.

Due to the direction of the winds and the fact that balloons could not really be steered, the stream of balloons went in only one direction—out of Paris. So, a later balloon, La Ville de Florence, transported carrier pigeons as well as mail. The pigeons were used by the French to carry messages back into Paris.

Since the balloons did not make their way back to Paris, the Parisians needed more and more balloons and began a flurry of balloon building. These new balloons were built with cheap materials and were often piloted by inexperienced aeronauts. Originating from the temporarily empty railroad stations and yards, they ferried people as well as mail and pigeons out of Paris. Some were barely able to reach a safe landing away from enemy lines. On October 7, 1870, the minister of the new French government, Léon Gambetta, made a dramatic escape from Paris by balloon, and with his chief assistant, Charles Louis de Saulces de Freycinet, established a provisional capital in the city of Tours.

By November 1870, the Prussians were reputed to have developed a special anti-balloon gun, and the French authorities ruled that balloons must leave Paris only by night. This added new hazards for the inexperienced aeronauts. Balloons could not be controlled, and they landed at unexpected locations, sometimes with fatal results when they landed in enemy territory. On one flight, two aeronauts became lost and drifted 800 miles (1,287 kilometers) to Norway. Two other balloons were lost without a trace.

Altogether, a total of sixty-six balloons left Paris during the siege, and fifty-eight landed safely. They carried some 102 people, more than 500 pigeons, and five dogs, which were supposed to return to Paris carrying messages but never reappeared. The balloons also delivered more than two million pieces of mail as far away as Tours, 125 miles to the southwest of Paris. And Janssen? He slipped out of Paris by night, sailing on one of the balloons that made it safely beyond the Prussian lines, on his way to view another eclipse of the sun. (He probably thought he was not so lucky, after all, for the day of the eclipse turned cloudy and he saw nothing.)

There are a few nonsensical reports floating around the internet which took seriously a story that he experimented further with helium, using his family and his dog as subjects for helium enemas—filling their intestines with this lighter-than-air gas—in an attempt to conquer flight without balloons. This story is, like so much internet "information," false (and perhaps intentionally funny), for although he had discovered the element, he had no idea of its properties, not even that it was a gas. In fact, all that was known until the end of the century was that the sun showed this one characteristic absorption line which could not be duplicated in laboratories on earth. In the same year as Janssen's discovery it was independently found by the English astronomer Norman Lockyer, who observed the same yellow line in the solar spectrum; it was Lockyer who gave it the name helium, after the Greek god of the sun. It took another twenty-five years before another English chemist, William Ramsay, confirmed the existence of helium by separating it from the mineral cleveite and determined its properties, the most important of which were its gaseous nature and nonreactive behavior—a complete understanding of which was not forthcoming until Niels Bohr's quantum theory.

Back at Brookhaven, helium seemed the way to start my mass spectrometer work; it was among the easiest of the noble gases to measure, and it was produced more abundantly in nuclear reactions than any other element. It has two stable isotopes, helium-4 and helium-3, which are produced in nuclear reactions in a ratio of roughly three to one. ("Natural" helium is composed of 99.9999% mass four and only .000137% mass three.)

The first problem I should have anticipated.[1] When an atom is smashed apart in a cyclotron, the resulting products are a mixture of stable and radioactive atoms. In my previous work, I had been detecting radioactive atoms by measuring their decays with a type of Geiger counter that sees the presence of individual atoms. But now I was preparing to search for stable noble gases with a mass spectrometer, which is a very sensitive instrument indeed—but not compared to a Geiger counter. The mass spec could detect and

1. When I graduated from college I thought I knew a good deal. When I finished my PhD I realized I was just beginning to learn.

accurately measure amounts of helium as small as 10^{-10} cubic centimeters (0.0000000001 cc), or roughly a few billion atoms, which is quite remarkable but is an enormous loss of sensitivity compared to my previous work. It meant we would have to bombard our targets in the cosmotron for at least six or seven hours (ten to twelve would be better) to produce enough helium to be measured.

And that turned out to be a problem. The entire crew of nuclear engineers who had built the cosmotron had gone to Europe that year to help build an even more powerful cyclotron at CERN, the European atomic energy laboratory, and had left behind a less experienced group who, as it turned out, weren't capable of keeping the machine running for that long at a stretch. Most nuclear physics experiments needed only a few minutes of running time, and this they could do, so these experiments were run during the day. Our work (Ollie Schaeffer's and mine) was scheduled at the end of the day.

So we would prepare our target and put it into the cosmotron at around five o'clock, the idea being that it would run throughout the night and we'd pick up the target in the morning. But it never worked out that way. I'd go home to my wife and daughter and have dinner—and get a phone call.

"We're having some trouble getting a beam," they'd say. "But don't worry, we'll keep trying."

So I'd play with the family and a few hours later get another phone call: "We've got a beam! Looking good." And eventually I'd go happily to bed, and then at two or three or maybe four o'clock the phone would ring again. "Sorry, we just lost the beam."

The worst part was that the target would have been irradiated for three or four or maybe five hours, and it might—it just might—be enough to give us measurable helium. So the next day I'd set up the mass spec and spend the day melting the target and purifying the gases and finally measuring the helium. And it was never quite enough to get a firm result. Day after day, week after week, one aborted run after another, and in the end we got nothing.

Finally Ollie suggested we try using meteorites as a substitute for the cosmotron.

ೋ

Meteorites are chunks of rock which fall to earth from space. At the time no one was quite sure where they came from, although the asteroid belt looked like a good bet. They range in size from dust to rocks to mountains or even moons—the two moons of Mars are a couple of asteroids which fell into Martian orbit. In fact, they're a good example of what we think happens. The asteroids are chunks of early solar system material that never conglomerated into a planet because of the stress induced by Jupiter's gravitational field (they lie between Mars and Jupiter). Whirling around out there, they continually bump into each other, and periodically one gets knocked out of its orbit. It can then wander off into space and fall onto Mars (or, just twice in solar system history, end up in orbit around Mars), or it might cross earth's orbit and so eventually collide with earth as a meteorite. More complex possibilities were being tossed around at the time but not seriously considered: an asteroid might hit Mars or the Moon[2] hard enough to knock a chunk off those bodies, which might then fall to earth as a meteorite, or meteorites might be cometary remnants. We didn't really know all that much about them.

One thing we did know was their age: 4.5 billion years.[3] For all that time they had been floating around in space and being zapped by cosmic rays, which are mainly high-energy protons originating in the sun and other stars. We didn't know much about the cosmic ray origins and fluxes, but we did know that their energy range was generally in the Mev to Gev region, the same energy range as the Cosmotron. Two earlier studies had shown that meteorites

2. "The Moon" is how we generally refer to our moon. "Moons" means the objects orbiting other planets; there are literally dozens of them. The terminology is unfortunate but historical: for thousands of years our moon was the only one we knew about.

3. Much more about this later.

contained helium produced by nuclear reactions induced by these cosmic rays during their passage through space. Ollie's idea was that if the cosmic ray flux was constant, the buildup of helium would be linear, and so we could measure the helium in the meteorites and thus estimate the time they had been exposed to the cosmic rays, that is, their lifetime in space.

Well, it was a little more complicated than that. For one thing, we didn't know what the cosmic ray flux was, or even if it was constant. But since the sun is a reasonably constant source of light (if it wasn't, life on earth would have been wiped out), as are most stars, it was reasonable to suppose that their cosmic rays were also produced at a constant rate. Estimating the rate was a different and more difficult proposition, but one that could be put off to begin with. For another thing, the idea was that the meteorites come from asteroids that are continually crashing into each other and knocking some out of asteroidal orbit and into an earth-crossing orbit, enabling them to land on earth. But this process would also change their size: they would be breaking apart as they crashed together, and this would influence how many cosmic rays actually react with them; an asteroid of more than a few meters would absorb the cosmic rays in its outer layers, effectively shielding the interior material. Then when it broke apart the inside would suddenly be exposed. When we pick up a meteorite on earth we don't know if it came from the inside of a large asteroid or from its outer shell, or even if it had ever been part of a larger object.

Still, Ollie reasoned, all these unknowns are part of what makes science fun. You make a start, and little by little you strip the unknowns away, and one day you realize you've discovered something new. It seemed like too complex a voyage for me, but I thought we could turn it around and use the meteorites to study the nuclear reactions I was interested in: instead of putting an iron target in the Cosmotron, we could use the iron meteorites that had been irradiated out in space. True, we didn't know the flux they had been exposed to, but we could at least get relative production rates of He-3 to He-4.

Wrong again. As it turned out, the many unknowns about meteorites made any unambiguous interpretation of the nuclear reactions impossible. The He-3/He-4 contents and ratios varied too much to lead us anywhere. But we were able to do what Ollie had suggested and use what we did know about nuclear reactions to learn something about the history of meteorites. For this, however, we needed to compare the helium to argon.

FOUR

Argon and the Rest

When the love-light is fading in your sweetheart's eye
Sail away—sail away . . .

—Noel Coward

THE DISCOVERY OF ARGON IS A GREAT EXAMPLE OF how little bitty precise measurements of stuff everyone knows sometimes lead to tremendous leaps of basic knowledge, because we don't always know what we think we know.

The story has a long and meandering lead-in, beginning with Aristotle's idea that "air" was one of the four earthly primeval elements, an idea which lasted some two thousand years—until the eighteenth century, when an Englishman named Joseph Priestley began to fool around in his homemade laboratory next door to a brewery.

Priestley by name, he was also priestly by nature. Educated as a dissenting minister, neither Church of England nor Roman Catholic, he taught and preached in a vigorously antiestablishment manner. Though he supported the American Revolution, he made no enemies in England for that, for there was as large a proportion of Englishmen as Americans who believed that the colonists were in the right. But when, a few years later, he sounded loud public hurrahs for the French Revolution, applauded the beheading of King Louis, and called for the same action against King George, he went a shout too far. Members of parliament called for action against the seditious minister, a mob broke into his home and ravaged it, and all in all he decided it might be time to sail away.

He was welcomed in America, where he was honored more for his theology than his scientific work (an opinion which he shared), until his thoughts evolved to the realization that the Christ story was simply an old superstition dressed up in Hebraic dress: God—Zeus, Wotan, Jupiter, Jehovah, take your pick—impregnates a human woman and the resulting child is less godly than the god-father but more so than the human mother. Jesus was only the most recent incarnation of the tale; Pythagoras, Alexander the Great, and a whole host of others share the same superstitious glory. But trying to convince his new compatriots was a losing game; his popularity waned, and the model community he planned was never populated.

Never mind. For us his importance lies in his chemical researches, which had been completed twenty years previously in Leeds. Like many of the early Christian theologians whose thoughts were concerned with the universe their God had created, he took a slight but ever-increasing interest in observing this natural world that God had wrought. He built a small laboratory and began by repeating experiments he had read about, but soon went on to conducting his own experiments.

Aristotle had died in 322 B.C. (one year after his most famous pupil, Alexander the Great), but his influence lived on well into the eighteenth century. (What a guy! He was wrong in just about everything, from the movements of the stars to the seat of human consciousness, but two thousand years later he was still The Man.) Air, as he taught, was still thought to be an indivisible substance, primeval in nature, but the brewery next door to Priestley's laboratory seemed to be producing it out of other materials. It had been known since antiquity that yeast stirred into a sugar solution soon created bubbles of "air," but no one before Priestley seems to have wondered at the fact. Could the air have been a component of the sugar, water, or yeast? Didn't seem likely. But the only other possibility was that the air was a combination of these substances, which would mean that it wasn't a basic facet of the universe—in other words, that Aristotle was wrong.

Curious, Priestley arranged with the friendly brewer to collect some of this air in bottles, and found to his surprise that a burning candle inserted into the brewer's air quickly went out!

Burning, that was the key. What a curious fact of the universe it is. Why do some substances, like wood, burn easily, while others, like iron or water, won't burn at all? To Priestley the answer was clear: like most other natural philosophers, he subscribed entirely to the phlogiston theory, in which this unseen, immeasurable substance, phlogiston, was the stuff of burning. If something had phlogiston in its essence, it burnt; if not, it didn't. The theory explained why something like wood could burn for a while, but then go out: when it had used up all its phlogiston, it was no longer burnable. It also explained why the burnt wood weighed less than the original: the loss in weight was the phlogiston which had been used up. Finally, it explained why a burnable substance enclosed with a fixed amount of air would stop burning sooner than if it were in the open air: when the enclosed air had taken up its fill of phlogiston, it couldn't accept any more, even though the wood had plenty to spare, and so no more could be emitted and the wood stopped burning.

An excellent theory, indeed, explaining all the observable facts. But then those damned natural philosophers did one experiment too many (a fault many of us still share): they burnt the metal magnesium, and found that it actually *gained* weight! Others began collecting all the soot and ashes that blew away in the smoke of a wood fire, and when they put this together with the burnt wood they found that not only hadn't the wood lost any weight at all, it too had gained weight.

What to do? One suggestion was that phlogiston had negative weight, but this was quickly ruled out as silly.[1] The next idea was that phlogiston did indeed have weight, but it was lighter than air

1. Not as silly as it sounds. Negative weight would mean some sort of negative gravity, and today we "know" that the universe is composed mostly of what we call dark energy, which seems to be exactly that: some sort of negative gravity. The more we learn about the universe, the more it seems that nothing is too silly to be true. (Except Creationism.)

so its loss would make the remains seem heavier. But quantitative experiments ruled this out. Nevertheless, Priestley saw no alternative and so expressed his results in line with the theory: his brewer's "fixed air," as he called it, simply had its fill of phlogiston, and so couldn't accept any more. This line of thinking, though false, nevertheless led him onward as he began to suspect the possibility of other types of air.

Using a magnifying lens, he focused sunlight on the compound of quicksilver that we know today to be mercuric oxide, and the heat produced a wonderful result. The emanated gas made flames burn even more brightly than did normal air: clearly, this meant that normal air contained a certain quantity of phlogiston which supported burning only until it reached its limit, while this new air must be "dephlogisticated" and so could accept more phlogiston than normal air, enabling things to burn more strongly.

Meanwhile, over in France, Antoine-Laurent Lavoisier was arguing that the entire concept of phlogiston was just a lot of hot air. By 1772 he showed conclusively that phosphorus and sulphur gained weight on burning, which he correctly attributed to their combining with air. Two years later, he extended his experiments to metals, correcting a mistake the English chemist Robert Boyle had made. Boyle had heated metals in sealed vessels and weighed them before and after. He explained the fact that they gained weight by hypothesizing that "particles of fire" (or phlogiston) passed through the glass and combined with the metals. Lavoisier, however, weighed the vessels—still sealed—with the metals inside them, and found that the total weight was the same before and after heating. He then broke the seal, allowing air to rush in. According to the phlogiston theory, the hot metal should then give up its phlogiston to the air, thus losing weight. But instead Lavoisier found that the metals gained weight.

This was the same year in which Priestley discovered his "dephlogisticated air," and when the two met in Paris to discuss their results, Lavoisier concluded that Priestley's dephlogisticated air was the stuff that combined with the metals to increase their weight, and was actually a new element, which he named oxygen. But somehow Priestley is given credit for its discovery, which doesn't seem quite fair, since he had no idea what it was he had discovered.[2] The fight for recognition, and for and against the phlogiston theory, grew bitter, and one can only wonder what Priestley's reaction must have been when finally the creatures of his idealized French Revolution ended the argument by cutting off the head of his antagonist.

Meanwhile, back in England, Joseph Black was carrying out further experiments on Priestley's "fixed air." By passing it over fine-grained charcoal he was able to absorb it—but never quite all of it. Some residual gas always remained, and this too was totally "phlogisticated," that is, it would not support a flame. Black's student, Daniel Rutherford, followed up by purifying the residual gas and showing that not only would a candle not burn in it, but a mouse could not live in it. Lavoisier later (though before his head was cut off) showed it to be still another element, which he called "azote," meaning without life, in homage to Rutherford's mouse experiment.

And now, a hundred years later, comes the little bitty precise measurement stuff which was to result in the discovery of all the noble gases. By this time, azote had been renamed nitrogen in English and had been recognized as the prime constituent of air. Techniques had been worked out to isolate it from air and to derive it chemically from other compounds, and it caught the interest of John William Strutt, the third Baron Rayleigh, whose researches had already covered a multitude of fields.

There are two opposite extremes of scientific types. At one end of the spectrum is the theorist who thinks large thoughts concerned with large-scale processes in the workings of the universe. At the other end is the practical experimentalist who spends his time tinkering in the laboratory, always trying to make the

2. Still, what comes around goes around: it was Priestley who first discovered Coulomb's law.

machinery just a little bit more precise. Most scientists fit somewhere between these two extremes, but seldom do we find someone who comprises both extremes in one body. Lord Rayleigh was such a man.

A sickly child with no apparent aptitude for any aspect of the intellectual life, he withdrew from both Eton and Harrow, was schooled at home in preparation for a life to be spent running the family's country estate, and as a last step in such preparation was sent to Cambridge, where to the astonishment and disapprobation of his family he sat for the Maths Tripos.[3] "Astonishment" becomes too small a word for their reaction when he bloomed under the tutorship of the mathematics faculty and emerged at graduation as the Senior Wrangler (top student). And "disapprobation" relegates itself to the infinitesimal compared to their feelings when he set himself up to follow a career in science, for as the eldest son he was the heir to the family title, and as such he had an obligation to fulfill the social responsibilities entailed as a noble paterfamilias.

Instead, he used his money to purchase scientific equipment, independence, and peace of mind, and he set off on a multitude of scientific researches that soon resulted in his being elected a Fellow of Trinity College (Cambridge)—a post which was taken from him when he decided to marry. (Nearly a hundred years later, a fellow professor of mine at Cornell, Henri Sack, dismissed a graduate student when the lad became engaged; he argued that it would take the boy's mind off his work.) With never a pause to regret the loss of prestige entailed, Strutt continued his work, at first on his estate and later at various university posts as his reputation grew sufficiently to outweigh the burden of being married. He was appointed the second Cavendish professor of experimental physics at Cambridge, following James Clerk Maxwell, and he transformed the teaching of physics there, introducing laboratory courses for the first time.

3. Students at the time concentrated exclusively on one subject. Tripos was the name given to whatever major they chose.

After five years he resigned, retreating again to his estate (with the death of his father he had become the third Lord Rayleigh) to concentrate on his researches, which led to further positions and honors. He was elected a Fellow of the Royal Society, ten years later received their Royal Medal, and was elected President of the Society in 1905. He was also President of the London Mathematical Society, and became chancellor of Cambridge University in 1908. He was the first person to explain why the sky is blue and sunsets are red (the short wavelengths of the sun's blue light are preferentially scattered in the atmosphere, thus coming to us from all over the sky, and at sunset the sun, low on the horizon, shines through more air and so loses even more of its blue light), and his other work ranged over both experimental and theoretical studies in optics, sound, waves, and electricity.[4] But what concerns us here is the research that led to the 1904 Nobel Prize, a few itty bitty measurements on the density of nitrogen.

At the time, Mendeleyev's periodic table was in an elementary state, with the recognition that there was a periodic relationship among the various elements but with no idea as to why. Rayleigh took up a previously noted point, that the atomic weights seemed to be nearly whole-number multiples of hydrogen—again without a decent theoretical explanation. He suggested that more careful determinations of the exact atomic weights might lead to further insights, and he undertook to do this himself, starting with the gaseous elements.

In general, experimental procedures are boring to everyone except those directly concerned, but here they hold the heart of the matter, for they lead to a totally unexpected result. As a friend of mine, Michael Lipschutz of Purdue University, is fond of reminding his students, obey the Biblical injunction to search out

4. He was also President of the Society for Psychical Research, although a disappointment to many of its members: in his Presidential Address of 1919, he stressed his lack of conviction for such stuff as communication with the dead and telepathy. See the *Proc. Soc. Psych. Res.* 30 (1919): 275–290.

knowledge, "Seek, and ye shall find," but he then adds, "But seek not to find that for which ye seek."

Rayleigh spent the next ten years on this seemingly routine task, which would have bored me to tears, measuring the precise atomic weights of hydrogen and oxygen—which yielded no important insights—before proceeding to nitrogen. To prepare pure nitrogen he began with air and a solution of ammonia (NH_3) . . . Well, let him tell it, as he did in his Nobel address: "Air bubbled through liquid ammonia is passed through a tube containing copper at a red heat where the oxygen of the air is consumed by the hydrogen of the ammonia, the excess of the ammonia being subsequently removed with sulphuric acid." That is, the air—a mixture of oxygen and nitrogen—was purified by removing the oxygen and reacting it with the hydrogen of the ammonia molecules, the red-hot copper acting as a catalyst. The resulting water vapor was condensed and the excess ammonia removed, leaving pure nitrogen, most of which had come from the initial air but with some addition from the ammonia.

He then determined the precise atomic weight of the nitrogen thus produced. The work was tedious, but he carefully completed it: "Having obtained a series of concordant observations on gas thus prepared I was at first disposed to consider the work on nitrogen as finished. . . ."

Most scientists would have done so, and moved on. Luckily, he didn't. Ever the careful, consummate experimentalist, he decided to check his results by going through the whole business again with a slight difference: instead of measuring the nitrogen which had been obtained from a combination of air and ammonia, he wanted to look at nitrogen from air alone. Working out a slight variation of his previous technique which eliminated the ammonia solution, he prepared purified nitrogen from nothing but air.

He anticipated an identical result from the atomic weight measurement (to be precise, he was measuring the densities of the gases, corresponding to their atomic weights), which would have confirmed his previous measurement. To his astonishment, he got something totally different.

Well, not exactly *totally* different: "to my surprise and disgust the densities obtained by the two methods differed by a thousandth part—a difference small in itself but entirely beyond the experimental errors." That is, repeated measurements on each of the two nitrogens agreed precisely with themselves, but differed from each other by one tenth of 1%. A lesser mortal (such as I) would have said, "Well, that's pretty good agreement," and gone home to my dinner. Rayleigh, however, said "It is a good rule in experimental work to seek to magnify a discrepancy when it first presents itself, rather than to follow the natural instinct of trying to get quit of it."[5]

The difference between the two nitrogens was in their preparation: one wholly from air, the other containing about 20% nitrogen from ammonia. So he repeated the ammonia preparation, but this time substituting pure oxygen for air, so that all the resulting nitrogen came from the ammonia solution, and indeed the discrepancy was increased: the ammonia-nitrogen was now 0.5% lighter than the air-nitrogen.

Since in the air case the nitrogen was known to be combined with no other element, while in the ammonia case it was chemically combined with hydrogen, the most obvious explanation was that although the gas in both cases was pure nitrogen, the two cases involved nitrogen in different chemical states. This line of reasoning went nowhere, however, and Raleigh thought further: "Among the explanations which suggested themselves were the presence of a gas heavier than nitrogen in air..."

He bent his efforts to isolating this hypothetical gas, and ultimately succeeded. To his surprise he found that it was impossible to make it combine with any other compound; it simply refused to enter into any chemical combination whatsoever, and so he named it argon, a word anglicized from the Greek word for "idle, doing no work."

5. A similar note was sounded by Sir Alexander Fleming when he cautioned a student to "always be sure to get everything possible out of your mistakes."

Similar experiments were being carried out at the same time by William Ramsay, professor of chemistry at University College, London. He and Rayleigh corresponded daily while working separately, and announced the discovery of argon jointly at the 1894 meeting of the British Association. A member of the audience, Sir Henry Miers of the British Museum, wrote to Ramsay the next day to tell him that he had heard of an experiment that seemed to be, though different, possibly related. An American chemist, William Hillebrand, who had received his degree at Heidelberg, had been working with uraninite, a uranium ore, and had noticed that when he dissolved it a gas was produced. Hillebrand thought it was nitrogen, and wasn't particularly interested in it.

Ramsay, thinking the gas might be argon, redid Hillebrand's experiment: the gas indeed was as inactive as argon but was much lighter. He sent a sample of the gas for identification to Sir William Crookes, who had equipment capable of measuring the spectral lines. To everyone's surprise, Crookes' measurements showed the yellow line characteristic of the element previously discovered by Janssen and Lockyer in the sun: the gas was not nitrogen, not argon, but helium.

The scepter passes now from Rayleigh to Ramsay, and from physics to chemistry—in particular, to the newly burgeoning periodic table and, as Ramsay humbly said in accepting his Nobel Prize, to his large, flat thumb.

He was a consummate experimentalist who rolled his own cigarettes, for the new store-bought ones were "unworthy of an experimentalist," and who modestly gloried in his thumb, for one of the major problems in purifying gases is how to move them around a vacuum tube without allowing any contaminating air to seep in. For his he used liquid mercury, and his "large flat thumb" was perfect for sealing off the ends of the eudiometer tubes that carried the mercury.

(Some fifty years later, at Brookhaven in the 1950s, we also used mercury, but instead of our thumbs we had mechanized pumps that would lower the mercury level to open the valves and raise it to close them. The problem was that the pumps had to be carefully temperature-controlled, and a mistake would send the mercury spilling out of the top and over the floor. All work then would stop until every drop of the viciously escaping quicksilver was chased down and vacuumed up, because the vapor emitted is a serious neurotoxin: Lewis Carroll's "Mad Hatter" was a familiar figure in bygone days, as hatters used mercury without realizing the danger. By the 1980s the mercury was replaced by mechanical leakproof valves—but they weren't always truly leakproof, and a good day's work was sometimes ruined when the analysis showed air contamination. Today you can put together a computer-operated system and sit back in comfort after pushing the "start" button and watch your gases being purified and measured. But of course computers sometimes have minds of their own, and The Glitch Who Stole Argon sometimes sneaks in to ruin a run. So, on reflection, perhaps Ramsay's flat thumb wasn't such a primitive tool after all.)

Just as important as his thumb was his recognition of what was then variously known to some as "the law of octaves" or "the periodic law" (or, to many, as an example of meaningless numerology). It had been noted as far back as the early nineteenth century, when Lavoisier had successfully argued that matter was composed not of earth, air, fire, and water but of various chemical elements; an Englishman, John Dalton, had followed by calculating the relative weights of these elements and their compounds; and a German chemist, Johann Döbereiner, noticed that several of the elements showed similar chemical habits and a consistency in atomic weights when organized into groups of three. His Law of Triads put calcium, strontium, and barium together in one group as similar chemicals, and the atomic weight of strontium (88) fell on the arithmetic mean between those of calcium (40) and barium (137).

Could this be coincidence? Yes, Döbereiner reasoned, if it was a singular correspondence, but not if it was repeated. Looking further, he recognized another triad of similar chemicals, lithium, sodium, and potassium, and once again there was the same relationship among their atomic weights: sodium (23) fell midway between lithium (7) and potassium (39). And there was also bromine (80), right between chlorine (36) and iodine (127).

By the middle of the nineteenth century we knew about sixty-two different elements, and when the English chemist John Newlands arranged them in order of increasing atomic weights, he noticed that a periodicity in similar chemical properties occurred every eight elements. As he put it, "the eighth element, starting from a given one, is a kind of repetition of the first, like the eighth note of an octave in music." In 1863 his Law of Octaves extended and replaced the Law of Triads. Finally, in the years 1869–1871, Dimitri Mendeleyev published his periodic table in much the same form as we see it today, albeit with only seventy-one elements instead of the hundred-plus we now know.

Of course, there was no Group 8 in his table: the noble gases hadn't yet been discovered. And now comes Ramsay's insight. "The discovery of argon raised the curiosity of Lord Rayleigh and myself as to its position in this table," he told the British Association in 1877. The two of them had recognized that in the periodic table there were several groups of three elements, such as iron, cobalt, and nickel, all of which had nearly identical chemical properties and atomic weights. And so, in this vein, "our first idea was that argon was probably a mixture of three gases... Indeed, their names were suggested, on this supposition, with patriotic bias, as Anglium, Scotium, and Hibernium!"

But with the discovery of helium not only as a solar element but as an earthly gas with, like argon, no propensity to interact, their reasoning took another tack. He and Rutherford had measured the atomic weight of the gas they called argon as 40, but since the other major constituents of air, oxygen and nitrogen, were both diatomic molecules, they had naturally assumed that argon also was, and hence its real atomic weight would be 20. But it was known at this time that the ratio of specific heat capacities of a gas is of two types, measured either at constant volume or at constant pressure, and when these measurements were carried out for argon the result was a ratio of 1.65.[6] A diatomic gas such as nitrogen or oxygen, with both rotational and translational energy, always shows values of 1.4 or less; only a monatomic gas (existing as single atoms and therefore with no rotational energy) can reach the higher value shown by argon, and so the results "had, at least in our opinion, unmistakably shown that it was molecularly monatomic, not diatomic, as at first conjectured, [and so] it was necessary to believe that its atomic weight was 40, and not 20." Similar measurements showed the same to be true for helium, giving an atomic weight of 4.

Since both helium and argon had chemical properties totally distinct from any other element (that is, they had no propensity for any chemical reactions at all), Ramsay proposed that they comprised a new set of elements; that is, a new column in the periodic table. And he went further, foreshadowing his proposition in the title of a talk he presented to the British Association in 1897: "An Undiscovered Gas."

He noted that in several other columns of similar elements the difference between the masses of neighbors is 16—fluorine (19) and chlorine (35), carbon (12) and silicon (28), oxygen (16) and sulfur (32)—and on this basis he predicted that there might be a new inert gas with mass 20 (4 + 16). This conclusion was buttressed by an error: manganese was wrongly assigned to the same chemical group as chlorine and fluorine, and the difference in mass between the heavier manganese (55) and chlorine (35) was 20. Similar mistakes led to a mass difference between the higher and lower elements sulfur (32) and chromium (52),

6. Specific heat is the heat required to raise the temperature of one gram by one degree C. The relationship between the two types was established in 1819 by the French chemists Pierre Louis Dulong and Alexis Thérèse Petit.

phosphorus and vanadium, and silicon and titanium, all of which showed the value 20. Thus the atomic weight of the new element, Ramsay reasoned, should be twenty less than argon, leading again to the value of 20. That is, argon minus 20 equals 20, as does helium plus 16.

"There should, therefore, be an undiscovered element between helium and argon, with an atomic weight 16 units higher than that of helium, and 20 units lower than that of argon, namely 20 . . . And pushing the analogy still further, it is to be expected that this element should be as indifferent to union with other elements as the two allied elements."

The problem of finding this new element seemed as difficult as that of finding the proverbial needle in a haystack. Even more so: they were to find a gas that wouldn't react with anything, that was invisible and odorless, that was present on earth in very low abundance (if it was on earth at all) and, if it did exist here, could be absolutely anywhere on this vast planet. As Ramsay put it, "here is a supposed gas, endowed no doubt with negative properties, and the whole world to find it in."

Well, not quite. While it's true that they had no inkling in what mineral it might be found, there is one place where an inert gas is likely to be: in the air itself. For if it doesn't react with anything, during the continual refluxing of the earth's crust it must be liberated from its hiding places and, once in the air, it's likely to stay there.

It took a while to reach that conclusion, and Ramsay reached it only reluctantly, because its abundance in the air must be very low, since no one had yet seen any evidence of it in all the experiments on the atmosphere that had been carried out in previous years. He hoped instead to find it concentrated in some mineral, but after two years of fruitless search he turned to what must have seemed his last hope. A London scientist, James Dewar, inventor of the vacuum flask, had managed to cool air down to the liquid state a few years earlier, and this, Ramsay decided, was the clue: gases of different atomic weight will have different boiling points.

So he first purified a vial of argon and then cooled it in liquid nitrogen until it too liquefied. What he had left in the vial was nearly—but not quite—a perfect vacuum: there was some residual gas still there. He passed an electric current through it, and it gave off a glaring crimson light: a light now familiar to all of us from shop windows and signs all over the world.

He had found his unknown gas, and he named it neon, from the Greek word for "new." And he went on. By analogy with other columns of the periodic table, he suspected there might be more gases of a similar stripe, heavier than argon, and in 1898 he was able to separate two more inert gases from the neon by repeated cooling. He named the lighter of the two krypton (again from the Greek, meaning "hidden"), and the heaviest gas he called xenon, "the stranger." Finally, two years later, together with Frederick Soddy, he completed the rostrum of noble gases by identifying an even heavier gas that had been found by others in thorium ores, and which they called radon because it seemed to be associated with the newly discovered phenomenon of radioactivity.

In 1904, when Rayleigh received the Nobel Prize in physics for his discovery of argon, Ramsay was awarded the chemistry prize for finding all the other noble gases.

You might not think this work was so important as to merit science's most important prize. After all, these gases are called rare because there is so very little of them on earth—with the exception of argon, which comprises merely 1% of the atmosphere, they are the rarest elements of all—and they are called noble because they do nothing at all. A group of elements that hardly exist and that do nothing should be worth nothing, shouldn't they?

And yet they have turned out to be of the utmost importance in understanding our universe, from the structure of the atom to the creation of the world. Consider, for example, the date of the world's creation. . . .

FIVE

Helium and the Age of the Earth

On this you can rely . . .
As time goes by.

—Herman Hupfeld

UNTIL NEARLY THE END OF THE NINETEENTH century, nobody was particularly interested in the age of the earth except a few theologians. In the second century A.D., the rabbi Yose ben Halafta wrote a tract known today as the Seder Olam (meaning Order of the World) in which he divided the history of the world into four parts: first, from the creation until the death of Moses; second, up to the murder of Zachariah; third, up to the destruction of the temple by Nebuchadnezzar, king of Babylon, in 586 B.C.; and finally, from then to his present day. The Bible gives the ages of the patriarchs at the time of the birth of their offspring: "This is the roll of Adam's descendants . . . When Adam was a hundred and thirty years old he became the father of Seth . . . When Seth was a hundred and five years old he became the father of Enosh . . ." So by adding the ages of the people listed in the Bible, ben Halafta calculated the passage of years in each period, concluding that the world was created 3,828 years before the destruction of the Second Temple by the Romans in 68 B.C. (an event now assigned to the year 70 B.C.); that is, creation took place in the year 3896 B.C. (3898 if we include the new date for the Second Temple).

There was little mention of his calculation until the Jews moved from Babylonia to Europe, and it then gradually came into use, replacing the then usual method of assigning dates as so many years after the beginning of the Seleucid era in 312 B.C. By the eleventh century it had been slightly revised so that the world was created in 3761 B.C., a date which became the basis of the Jewish calendar; as I write this (2009) we are in the year 5770 A.M., or Anno Mundi.

No more was heard about the problem until, in a brief work entitled *A Few and New Observations upon the Book of Genesis* (London, 1642), the Anglican bishop John Lightfoot, vice-chancellor of the University of Cambridge, dissected Genesis verse by verse and came up with an astounding result: the entire universe, including our world, was created on Sunday, September 12, 3928 B.C., and man was created on Friday, September 17, 3928, at 9 A.M.

How did he arrive at this precision? Easy; all he had to do was take the Bible as the literal word of God.

So then: "In the beginning God created the heavens and the earth, and all the host of them." So it is clear that "Heaven and earth, centre and circumference, were created together in the same instant." The year of creation was found, as the old rabbi did, by counting the ages of people in the Bible. Next, Lightfoot decided that this moment of creation was of necessity at either the vernal or the autumnal equinox, because these are the two times of year when the day is divided equally between light and darkness, and he thought that was reasonable. He chose the autumnal equinox for reasons that he never clearly explained, and he calculated that it fell on September 15 in the year 3928 B.C. Today we know that the equinox that year was September 21. So okay, he was a few days off because he didn't know that it changes on a thousand-year time scale; surely we can forgive him for that.

But then, catastrophe! His calendar showed that September 15, 3928 B.C. was a Wednesday, and the date of creation had to fall on a Sunday, without question, because that's the first day, God having rested on the seventh day, the Sabbath. What to do?

After searching his soul he concluded that the prime necessity was the day of the week, Sunday, and perhaps an error had crept into his calendar. The closest Sunday to his calculated equinox was the twelfth of September, so he settled on that as the fateful day. Finally, man was created on the following Friday (the Bible clearly stating that this took place on the sixth day), at "the third hour of the day, or nine of the clock in the morning"... Well, God only knows how he arrived at this.

A decade later James Ussher, Archbishop of Armagh and Primate of All Ireland, jumped into the game. A staunch Catholic in an Ireland ruled despotically by the godless English Protestants, he argued vociferously for the supremacy of the old religion: only Catholics could understand the revealed Word of God. To buttress his claim, he determined to seek out the date of Creation for himself: "I judge it indeed difficult but not impossible... to attain, not only the number of years, but even of days from the Creation of the World." He recalculated the death of Nebuchadnezzar as occurring 3,442 years after the creation of the world; history records that he died in 562 B.C.; 3,442 plus 562 equals 4004 B.C., so that's when the creation happened. In that year, the Sunday nearest to the autumnal equinox (from Lightfoot's calculation) was October 23. From all this he deduced that the creation of Heaven and Earth "fell upon the entrance of the night preceding the twenty third day of *Octob.* in the year of the Julian calendar, 710," which was, of course, a Sunday. (The expulsion from Eden occurred barely two weeks later on Monday, November 10, and Noah's flood ended on Wednesday, May 5, 1491 B.C.)

Ussher died in 1656, but his dates lived on. In 1675, Thomas Guy (later Sir Thomas, the founder of Guy's Hospital in London) was a businessman who branched out into publishing. He concluded an arrangement with Oxford University to publish a new edition of the Bible with their imprint, and had the happy idea of inserting Ussher's dates in the margin alongside the events described.

The version gained such popularity that in 1701 the Church of England adopted it, and thus the dates became inscribed as an "authorized" commentary in the King James Bible, so that people began to accept it as an authentic part of the original Bible (whatever that is).

It seemed a pretty reasonable number at the time; at least, no one attacked it seriously, although there was a Scots geologist, James Hutton, who in 1785 concluded from his studies that the earth showed "no vestige of a beginning, no prospect of an end." But that was vague enough to be generally ignored, and besides, until we began to search the planet for oil no one took geology seriously. It was a hobby for dilettantes; Hutton himself was a farmer and physician.

As we moved into the eighteenth and nineteenth centuries, astronomy was the science of the day, and the astronomers showed little concern about the date of creation. The only contribution they could make was to discuss the energy of the sun. If the sun's fire consisted of the same sort of fires we saw on earth, the oxidation of carbon, then from the rough numbers they had on the carbon and oxygen content of the sun combined with the observed energy output, they concluded that the sun might have a lifetime of roughly ten thousand years, and that fit well with the Biblical estimate: the "heavens and the earth" were created six thousand years ago, the sun would burn another four thousand, and at that time—or perhaps sometime before that—Jesus would return and accompany us all to Heaven.

So everyone was pretty much happy. And then that insensitive bastard Charles Darwin came along.

Darwin started off, like many of us, as a religious young man. No one on His Majesty's Ship Beagle could outquote him on Biblical issues. But as he matured and learned more about this world of ours, as he sought to reconcile his discoveries with his God and failed, his faith faltered, lessened, and turned into contempt. The Bible was "no more to be trusted than the sacred books of the Hindoos" [sic] or the beliefs of any other barbarian.

And the theologians thought the same of his ideas, with seemingly as much reason, for throughout history we have not seen creatures evolve. Elephants were elephants in Hannibal's time, fish were fish in Homer's, locusts were locusts when they rained down on Pharaoh some three thousand years ago. Darwin was forced to admit that his theory needed a very old earth, many millions of years old at the very least. His own studies on geology led him to suggest an age of some thirty million years, but this was attacked on geologic grounds by those who, it seemed, knew more of the subject than he, and he retreated to calling for merely "a very long time."

His theory seemed so reasonable on all other grounds that it sparked an interest among scientists. Was there nothing science could contribute to the problem of the age of the earth?

As it turned out, there was.

The shape of the earth gave the first clue. Isaac Newton recognized that gravity (equal in all directions) combined with the centrifugal force of the spinning earth (directed outward along the equator) would result in the observed spherical shape with an equatorial bulge, if the earth had formed in a molten state. The next step was to estimate how long this molten mass would take to cool down, and this was undertaken by George-Louis Leclerc, Comte de Buffon, in 1778.

Buffon made up a group of ten iron spheres varying in diameter, with a maximum size of five inches. He heated them to a white heat and then measured how long they took to cool, finding that the cooling time was a linear function of their diameter. He then extrapolated his results to an iron sphere the size of the earth, and calculated that it would take 96,670 years to cool down to the earth's present temperature.

He realized that the precise number was only an approximation, and a bad one at that. From what he knew of geology he thought the earth was actually much older, but no matter how he varied the sizes and compositions of his spheres he got much the same value: some tens of thousands of years.

Which pleased nobody. His results weren't long enough for Darwin nor short enough for Ussher, but his method pointed the way to the future, and nearly a hundred years later William Thomson, Lord Kelvin, took up the struggle.

Kelvin presents a conundrum: not only was he the preeminent scientist of his day, but at the same time he was the preeminent scientific reactionary. In science as in politics one often finds people liberal in their youth and conservative later in life,[1] but Kelvin was only in his thirties when Darwin's theory burst on the scene and violated the basic tenets of his soul. Descended from apes? Nothing but animals ourselves? Bloody nonsense! But not being a biologist himself, he had sense enough not to attack the theory on biological grounds. Unfortunately, he didn't have sense enough at least to look at the biological data Darwin had accumulated; he simply ignored all that. Instead, he concentrated his energies on attacking Darwin's silly but necessary hypothesis that the world was hundreds of millions of years old. Here was a field that physics could enter, and thereby dispel the evolutionary miasma that was clouding our vision of the truth.

It was the age of coal, and many of the British aristocracy made their fortunes in the coal mines. (Not working in the coal mines themselves, of course, but working the rich veins with miners who were paid poverty wages and suffered dangerous, slavelike conditions—but that's another book.) At any rate, evincing an interest in the mines, Thomson—or Lord Kelvin, as he became—made several amusing trips down into the tunnels and realized that, going deep into the earth and away from the warming sun, it didn't get colder down there but warmer.

1. "If you're not a liberal when you're young, you have no heart; if you're not a conservative when you're old you have no brain." Or so such folk like to think.

Of course! How do volcanoes erupt, after all? The earth must be hotter inside than out. Which means it must be cooling, with heat constantly flowing out into space. Kelvin was the first to formulate the principle that later would become the basis of the Second Law of Thermodynamics: heat will always flow from hot to cold, in this case from the hot interior of the earth to the cold vastness of space. And so, knowing something about the heat capacities of the materials that make up the earth, one should be able to calculate precisely how long it has taken the earth to cool to its present state.

Beginning in 1862 he published a series of papers on the subject, with apparently increasing precision as he took note of more and more experimental facts, beginning with an estimate of "somewhere between 20 and 40 million years," and ending with a rather firm number of 24.6 million—which of course pleased neither faction, being (like Buffon's earlier estimate} too old for the Biblical purists and yet too young for Darwin.

A different approach was taken by the Irish geologist John Joly, who began by wondering about a totally different question ("Seek and ye shall find, but seek not to find that for which ye seek"). Concerned more with the earth as it is today than with its creation, he was asking himself why the oceans are salty while lakes and rivers are fresh, and he came up with the answer, which he based on the hydrologic cycle.

To begin with, back in the early days of the earth, all water is fresh as it comes bubbling up from the interior. Great freshwater oceans form over the low-lying portions of the earth, and as the waters sit there they are warmed by the sun. Being warmed, their surfaces evaporate. The water vapor rises and cools, forms clouds, and rains back down again. The rains that fall back into the sea whence they arose cause no change, but some of the clouds drift over the continents, and there the falling rain forms lakes, which overflow into rivers, which curl around and carve their way through the rock and find their way eventually back into the oceans. As they work their way through and over the earth, they dissolve the terrestrial minerals, for water is a most corrosive liquid, carving deep valleys out of mountains and bringing the dissolved minerals ("salts") into the oceans.

As time goes on and the cycle continues, the oceans gradually become saltier and saltier, for though the water continually evaporates and goes through the cycle, the salts do not; they remain in the oceans, building up over time.

And suddenly it becomes possible to calculate the age of the earth. All we have to do is measure the yearly inflow of salt and the total salt content of the oceans today.

The latter is easy enough, but the former has many obstacles. Each of the world's rivers flows through different regions with different minerals, and carries different amounts of water. Even if you could measure every river in the world, how could you account for possible secular variations: as the earth is worn away, wouldn't the salt contents of the rivers change with time? Of course they would, and despite the complications Joly set himself the task of measuring that time. From what he knew of geology and chemistry, he made estimates of the various parameters and arrived at the conclusion that it had taken 90 million years for the oceans to reach their current level of saltiness.

Ridiculous, Kelvin fumed. He didn't bother to attack the geological argument; as with Darwin's biology, he just ignored it. Physics, having proved the earth couldn't be more than 25 million years old, was far above such plebian sciences as geology. The geological argument simply had to be wrong.

As it turned out, he was right about geology being wrong but wrong about physics being right. Joly was wrong about the salting rate of the oceans, and Kelvin was wrong about the cooling rate of the earth, and so at last we come back to helium.

But not quite yet. First we travel to New Zealand, where in the year 1860 the patience of the native Maori finally broke and a fierce war broke out in North Island between them and the immigrant English. The Shuttleworths of Sussex took their daughter Caroline, aka the widow Thompson, and her four children to South Island to find safety. Six years later, one of Caroline's daughters, Martha, became the local schoolteacher and married a young farmer, the son of a wheelwright from Dundee. The farmer's name was James Rutherford (no known relation to Daniel Rutherford of the previous chapter), and their fourth child (of a total brood of twelve), a brawny bear cub of a boy named Ernest, now enters upon the scene.

His parents couldn't afford a college education for him, but the local college offered an examination-based scholarship. Determined to "make something of himself," he took the exam—and failed. But he sat for it again a year later, and this time he succeeded. After graduating he wanted to teach, but there were no openings, so he eked out a bare living tutoring. He was saved from this drudgery when England established a set of research scholarships alternating among the graduates of universities outside of the home country; every other year, one such was to be granted to New Zealand. This was his last chance to break out into the larger world, and immediately he applied, as did one other person.

The other person got it.

But the winner decided not to take it. By default, then, it went to him, and finally Ernest Rutherford left New Zealand for the real world of science at Cambridge University's Cavendish Laboratory. He began his researches there on electrical conduction (his mentor was J. J. Thomson, who had just discovered the electron and had formulated the model of the atom which Ernest was to destroy), but he switched to radioactivity as soon as it was discovered, in 1896. Two years later, he found that there were two types of radioactivity, which he named alpha and beta. The betas were soon identified as Thomson's electrons; the alphas are where this story is going.

Under the terms of his scholarship, which had been designed to train colonials to work in the colonies, Rutherford wasn't eligible for a permanent position in England, and so he accepted an offer from McGill University in Canada. He found it to be a better place than he anticipated, with well-stocked laboratories and well-educated colleagues, among whom was a young chemist also interested in this new business of radioactivity. Frederick Soddy turned out to be the key with which Rutherford would unlock the mysteries of the atom. What they had on their hands—what Madame Curie and her husband had found—was a totally mysterious outpouring of energy with no observable physical change. That is, the Curies' new element, radium, just sat there pouring out energy forever. This was impossible. The first law of thermodynamics, as every student learns (although not necessarily learning that name), is that neither matter nor energy can be created (nor destroyed), yet here was energy seemingly being created, endlessly and forever. Pierre Curie used to astound his lecture audiences by placing a block of ice on the podium at the beginning of his talk, taking out a vial of radium powder from his vest pocket and placing it on the ice. He then proceeded to give his lecture, during which the audience would see the vial slowly melting the ice around it and sinking further and further into the block, disappearing before their eyes. At the end of the lecture he would retrieve the vial, hold it aloft to show that nothing in it had changed, and invite members of the audience to touch it to see that it was still warm. He would then put the vial back in his pocket, where he kept it. When a rash developed on his skin he merely switched pockets. Radioactivity, his widow would later declare, was a cure for cancer, not a cause. (She was both right and wrong in this, and it is not clear that Pierre, who died by falling unconscious under the wheels of a carriage, ever did develop cancer.)

By 1902 Soddy and Rutherford had found that indeed something *was* changing in the radium; it was transmogrifying into another element! This was astounding, and to many scientists of the day it was unforgivable. What S&R were claiming was nothing less than a return to the old and discredited claims of alchemy, the transmutation of the elements. What next, changing lead into gold? Since the triumph of atomic theory it was known by all true scientists that the chemical elements were the basic constituents of matter, unchanging and everlasting—but what everyone knows is not always true.

Soddy and Rutherford were right; the energy coming out of radium was due to its transformation into a series of different elements, ending in lead. (These chemical changes were difficult to measure, and hadn't been observed before, because so very few of the atoms are actually changing.) Because they were getting an observable amount of energy released without an observable number of atoms changing, it was clear that the energy involved (the energy per atom) was tremendous. That is, in a normal fire you get a lot of heat energy coming out, but you also see a large chemical change: the wood turns to ash before your eyes. This meant that if you could get a large number of radioactive atoms to give off their energy, the amount of energy would be unbelievable. But, Rutherford warned, forget it: "The energy produced by the atom is a very poor kind of thing. Anyone who expects a source of power from the transformation of these atoms is talking moonshine."

Oh well, as Joe E. Lewis concluded in *Some Like It Hot*, nobody's perfect.

But if he wasn't perfect, he was pretty damn good. The change of elements and the heat produced were one clue; the next was the rate of the reactions. By chemically separating the various radioactivities from uranium minerals, Rutherford and Soddy showed that the activity was, first of all, proportional to the amount of material present and did not depend on chemical state, temperature, pressure, or anything else at all. This meant that the activity A was simply proportional to the number of atoms N; changing the proportionality to an equation by inserting a constant, q, they arrived at

$$A = qN \tag{1}$$

Recognizing that the radioactivity A was due to a change in the number of atoms, using calculus they now had

$$dN/dt = qN \qquad (2)$$

Integrating this they arrived at

$$N = N_0 e^{-qt} \qquad (3)$$

In other words, the number of atoms, and therefore the radioactivity, should decrease exponentially with time. And indeed, when they followed the various activities over time, they found that this was so.

Finally, they wondered about the association of helium with uranium. Remember, helium was always found in uranium (or other radioactive) minerals, and nowhere else. The uranium (and its associated radium) was changing into other atoms, but they couldn't be changing into helium, could they? No, not really, for uranium and radium are among the most massive atoms, and helium is (except for hydrogen) the smallest. But then they noticed that the atoms formed by radioactivity seemed often to have an atomic weight four less than their parent, and helium has exactly that mass. Perhaps, they conjectured, the alpha rays were actually particles[2] of helium.

From conjecture to experiment was the work of a moment. They had their glassblower create a flask with extremely thin walls, and they pumped the atmosphere out of it. They placed an alpha-radioactive sample next to it, so that the alpha "rays" might pass right through the glass into the flask. After only a few hours, they found the characteristic spectral line of helium within the flask, and after a few days it had grown in brilliance so greatly as to be unmistakable.

Finally, putting all this together, Rutherford measured the age of the earth. First, however, he had to explain the error in Kelvin's work. (He didn't bother with Joly's estimate; Joly, remember, being only a geologist, was beneath the notice of a physicist.) Kelvin had

2. The concept of an atomic nucleus had not yet been thought of; thus the indefinite "particle."

based his age on the cooling of the earth, and now Rutherford knew that he was wrong because the earth was not simply a cooling body: it had in its innards large quantities of radioactive elements that were constantly producing heat. You might calculate, for instance, how long it would take a turned-off oven to cool. But even if you knew everything about its insulation, if someone had left the oven on your answer would be wrong. And that is what had happened; someone had left the earth's oven on.

In 1904 Rutherford, by now a Fellow of the Royal Societies of both Canada and England, was invited to give a talk at London's Royal Institution and was appalled to find Lord Kelvin sitting in the front row. Rutherford may have been a brash young colonial, but he was neither brash enough nor young enough to contemplate without terror his intention of insulting the most revered scientist in the world—for an insult it would surely be, to state that the old man was wrong, that the earth must be much older than his estimate of some 20 million years.

But what could he do? He had no doubt that the radioactive heat generated within the earth was sufficient to keep it warm long beyond Kelvin's estimate. He began with an overview of the situation, and with relief he saw Kelvin fall asleep in his chair as he talked. But Kelvin's apparent somnolence was not quite a coma, for just as Rutherford began to break the news of the excessive heat and the consequent longevity of the earth, "I saw the old bird sit up, open an eye and cock a baleful glance at me! Then a sudden inspiration came, and I said Lord Kelvin had limited the age of the earth provided that no new source of heat was discovered! That prophetic utterance refers to what we are now considering tonight. Behold! The old boy beamed upon me."

He didn't beam for long. To the end of his life Kelvin never accepted the importance of radioactivity. Soon after Rutherford's talk the two of them were invited to Lord Rayleigh's home for a weekend, and Rutherford wrote to his wife: "Lord Kelvin talks [about radioactivity] much of the day, and I admire his confidence in talking about a subject of which he has taken the trouble to learn so little...He won't listen to my views...but Strutt [Rayleigh] gives him a year to change his mind. In fact they placed a bet to that effect." Rayleigh lost the bet. To the end of his life Kelvin argued against the reality of nuclear energy, blaming the experimental observations on an aether that doesn't exist.

So the earth was older than 20 million years, but by how much? It was another two years before Rutherford was able to put together his work on radioactivity to begin to get an answer. If the number of uranium atoms in a mineral was declining every year and the number of helium atoms was increasing, then the ratio of uranium to helium would be changing every year; that is, the ratio would be a measure of the age of the mineral. To accomplish this he had first to use equation (1) to determine the constant q by measuring the mass of uranium in a sample[3] and at the same time its radioactivity, and then simply[4] measure the amounts of uranium and helium in a mineral. By 1906 he had done this and found a range of values, which was to be expected for two reasons.

First, helium is a small gaseous atom and might easily leak out of a rock, especially over periods of millions of years during which the rock might have been subjected to varying temperatures. But more importantly, rocks are forming all the time (i.e., from volcanic eruptions); you can't expect any rock you pick up to have been created during the formation of the earth. Consequently the earth must be at least as old as the oldest rock, and in 1906 he published values of roughly 500 million years for two uranium minerals. The earth was at least hundreds of millions of years old, and Darwin breathed a sigh of relief (and the theologians a snort of disgust).

3. The number of atoms is the weight divided by the atomic weight and multiplied by Avogadro's number (6×10^{23}).

4. Okay, it's really not that simple; we're skipping over a few complications that don't change anything. For a full discussion, see Dalrymple.

In 1908 Rutherford was awarded the Nobel Prize—not in physics but in chemistry, "for his investigations into the . . . chemistry of radioactive substances." He liked to tell his friends that of all the (radioactive) transformations he knew, the fastest was his own transformation from a physicist to a chemist.

Other people quickly took up the work, and this technique of radioactive dating continues unabated today. It was soon discovered that argon is the result of the radioactive decay of potassium, and since argon is a larger atom than helium, it is less likely to leak out. Today potassium-argon is a standard tool of the geochronologist (particularly in its modern mode, known as Ar-Ar dating; see chapter 14). More reliable, under certain circumstances, is the decay of uranium to its final product, lead, and in 1953 a Caltech scientist, Claire Patterson,[5] obtained a firm date for the earth of 4.5 billion years, a date which remains essentially unchanged today.

Even more astounding is another rare gas, xenon, with which we can date—But I'm getting ahead of myself. Before we talk about that, let's discuss . . .

5. I once roused an audience to laughter by ignorantly referring to him, whom I had not yet met, as "her."

SIX

The Strange Case of Helium and the Nuclear Atom

It was quite the most incredible event that has ever happened to me in my life.

—Ernest Rutherford

IT IS OFTEN TAKEN AS A MATTER OF ESTABLISHED fact that the difference between a good scientist and a great scientist is the ability to distinguish in advance which problems are going to be the important ones. I think this belief is a reflection of the fact that history is written by the winners: Professor X chooses a problem and with much hard work solves it, but it turns out not to have important consequences, so it and he are forgotten; Professor Y does the same, but this time the result spurs further work or even opens new and unforeseeable regions of science, so he naturally feels that his "intuition" was correct. But how do you distinguish his intuition from a lucky guess?

I suggest that a study of the history of science tells us that luck plays a significant part. Consider, for example, Lord Rutherford's discovery of the nuclear atom—perhaps the most important experimental discovery of the twentieth century, in that it led to quantum theory and the whole of nuclear physics.[1]

To set the stage:

By the first few years of the twentieth century it had been determined that there were three kinds of radioactive emissions, termed alpha, beta, and gamma rays. The gamma rays were electromagnetic in nature, the beta rays were electrons, and Rutherford had just shown that the alpha rays were in fact helium; or rather, as he put it, the alpha rays were a stream of particles zipping along at roughly 10,000 miles per second which, after they slowed down and lost their electric charge, became helium atoms. (He didn't realize at the time that they "lost" their positive electric charge by picking up negatively charged electrons.)

What next?

Well, the natural thing to do was to see how these radioactive emissions interacted with matter. This had already been done with the beta and gamma radiations: a stream of these radiations had been directed at various targets, and such parameters as their depth of penetration and ionizing capabilities had been measured, with no particular insights gained (an example of Professor X's work). Still, it was the sort of scut research that filled in the blanks and, more importantly, was useful in training new scientists in the methods of research. Thus, when Rutherford's assistant, Hans Geiger (later inventor of the Geiger tube), came to him and said... well, this is how Rutherford himself put it:

> One day Geiger came to me and said, "Don't you think that young Marsden,[2] whom I am training in radioactive methods, ought to begin a small research?" Now I had thought that too,

1. For the 2010 Academy Awards, Dave Barry wrote this joke (which was cut due to time constraints): "*Up* [winner of best documentary] tells the moving story of a bitter old man whose life is transformed by the most powerful force of all: helium." Though he didn't realize it, to a very real extent Mr. Barry was not joking.

2. Ernest Marsden, a twenty-year-old undergraduate, who later became a first-rate scientist.

> so I said, "Why not let him see if any α-particles can be scattered through a large angle?"

Geiger had already been looking at the interaction of the alphas with matter, in particular at how they scattered from metal foils. The current theory of atomic structure, promulgated by Rutherford's mentor, J. J. Thomson, was that the atoms were a sort of mushy pudding composed of protons and electrons, with empty spaces between. If a collimated stream of alphas was directed at a very thin foil, say just one atomic layer thick, some of the alphas would pass through the empty spaces and some would blast their way through the mushy atoms—remember, the alphas are zooming along at ten thousand miles a second, so they would burst right through with only a light bump in their direction; that is, they would be scattered through a slight angle. In the more realistic case of a foil many atomic layers thick, the alphas would hit multiple atoms on their way through, each of which would contribute a small amount to the scattering. But even in this case of multiple impacts the percentage scattered would die away quickly with each increase in angle.

What Rutherford was now suggesting was that Marsden take up the experiment, looking at large angles to see if any alphas were scattered very much. According to the accepted theory, none would be. But it would be a good way for a young student to get started on experimental work. Rutherford continued:

> I may tell you in confidence that I did not believe that they [alphas] would be [scattered at large angles], since we knew that the α-particle was a very fast, massive particle, with a great deal of energy, and you could show that if the scattering was due to the accumulated effect of a number of small scatterings the chance of an α-particle's being scattered backwards was very small. Then I remember two or three days later Geiger coming to me in great excitement and saying, "We have been able to get some of the α-particles coming backwards. . . ." It was quite the most incredible event that has ever happened to me in my life. It was almost

> as incredible as if you fired a 15-inch shell at a piece of tissue paper and it came back and hit you.

This is the way science often works, particularly in the case of great advances. It's called serendipity, a word coined by Horace Walpole in 1754 on the basis of "a silly fairy tale, called the three Princes of Serendip," in which the mythical princes "were always making discoveries, by accidents and sagacity, of things which they were not in quest of." Someone not as blunt and honest as Rutherford might have said, "I had an intuition that the current theory of atoms was wrong, and was searching for proof." But to good old honest Ernest it was admittedly a bolt from the blue, "the most incredible event that has ever happened to me in my life."

What is perhaps even more incredible is how he picked up this loose ball and ran with it. How was it possible for this massive and energetic alpha particle to come bouncing backwards? J. J. Thomson's idea of a mushy atom simply couldn't account for it, and so that idea had to go because of the basic foundation of science.

Well, actually there are two such basics. Theorists often focus on beauty: the way our universe happens to be constructed, if a theory is beautiful it often turns out to be true. But beauty is in the mind's eye of the theorist, and so sometimes this argument fails. In the 1950s, for example, I and a few others thought that the Gold-Bondi-Hoyle steady-state theory of the universe was beautiful and the Big Bang was messy, but steady-state turned out to be just plain wrong. Which brings us to the other basic foundation of science: actual observations of the universe. Seeing, after all, is believing.

Take the bumblebee, for example. You will sometimes read that scientists believe the bumblebee can't fly because its wing structure doesn't fit our theory of how things fly. But such a statement is total nonsense: clearly the bumblebee does fly, and no scientist would hold to a theory which disputes an observational fact.

BUT. When is an observational fact truly a fact? Observations, no less than theories, are subject to error. For example, within the past several years no fewer than a hundred men who had been convicted and sentenced to death largely on the strength of the testimony of witnesses—that is, on the basis of observations—have been proven innocent by DNA testing. The observations were simply wrong. And in science, the first experimental test of Einstein's theory of relativity "proved" that it was wrong. Einstein agreed that if the experiment was right the theory had to be wrong, but he argued that the theory was so beautiful he thought the experiment would turn out to be wrong, and, upon repetition by others, so it did.

In Rutherford's case his group put together such immaculate observations that everyone was convinced, but then his conclusions came into conflict with a more beautiful theory than Thomson's, and the world was never again the same.

The problem was that if Thomson's pudding theory of the atom was correct, "a simple calculation," as Geiger put it in a paper published in 1910, "shows that the probability of an α-particle scattered through an angle exceeding 90° is extremely small and of a different order from that which the reflection experiment suggests." In other words, Thomson's atom could not bounce the alphas backwards as the experiment showed it did. Geiger went on to say, "It does not appear profitable at present to discuss the assumption which might be made to account for this difference." In other words, neither he nor anyone in Rutherford's lab had any idea how to account for the experimental results.

This impasse lasted a year and a half, during which Rutherford fumed and fretted until finally, in December of 1910, he burst into Geiger's room and announced that he finally "knew what the atom looked like and how to explain the large deflections of the alpha-particles." He knew that the alpha particle was a positively charged helium atom, and surmised that the large scatter was the result of the electrostatic repulsion between its positive charge and that of the target atoms. "Supposing that the forces involved in such collisions are of the ordinary electrostatic type," he reminisced in a lecture a dozen years later, "it can readily be calculated that in order to produce such a large deflection of the alpha particle in an atomic encounter, the atom must contain a massive charged centre of very minute dimensions. From this arose the conception of the now well known nucleus atom, where the atom is taken to consist of a minute positively charged nucleus containing most of the mass of the atom, surrounded at relatively great distances by a distribution of electrons equal in number to the units of resultant positive charge on its nucleus."

Unfortunately, this led to a more disturbing problem than it solved, for this explanation of the nature of the atom was in serious conflict with a truly beautiful theory: Maxwell's theory of electromagnetism.

In a series of papers extending from the 1850s to the 1870s, James Clerk Maxwell had united a number of observations on electricity and magnetism and explained them with his electromagnetic theory. Einstein, among others, thought the theory was just too beautiful not to be true. (It was one of the factors which led him to relativity: it explained light as an electromagnetic wave with a constant speed, and when Einstein asked himself, "A speed constant relative to what?" he was on his way.)

The problem that Rutherford encountered was this: if the atom consisted of a positive nucleus surrounded by negative electrons, what kept the electrons from being pulled into the nucleus by their mutual electrostatic attraction? The answer was clear: the electrons couldn't be simply sitting there in space but had to be whirling around the nucleus so that an inertial (centrifugal) force would be set up to balance the electrostatic attraction. But here's the rub: Maxwell's theory—too beautiful not to be true, remember—insisted that electromagnetic energy-carrying waves are emitted from any accelerating charged particle, and if the charged electrons are whirling around in circles, they are constantly changing direction and therefore accelerating[3] and therefore emitting electromagnetic waves and therefore losing energy and therefore they must inevitably spiral into the nucleus and Rutherford's atoms must self-destruct. A simple calculation showed this would happen within seconds. In other words, Rutherford's atoms could not exist!

Obviously, either Rutherford or Maxwell had to be wrong. Until a young Danish physicist, Niels Bohr, came along and made a truly outrageous suggestion.

Bohr's doctoral thesis in Copenhagen dealt with the then rudimentary knowledge of electrons, and so he came to Cambridge for further study with the man who had discovered electrons, J. J. Thomson. But Bohr's manner was shy, and this, exacerbated by his poor English, made meaningful communication with the introverted Thomson difficult. (I heard Bohr lecture once. Or rather, I *saw* him lecture; I heard very little and understood absolutely nothing. His head was bowed and turned almost obsessively to the blackboard, on which he wrote tiny, barely legible mathematical symbols, and he whispered to the blackboard in a thick Danish accent. There is a story about him which, if it isn't true, ought to be. During a lecture he wrote an equation on the blackboard, whispered "From this it obviously follows that..." and wrote another equation. Then he stopped, stood silently gazing at what he had written, started walking back and forth, his head bowed, muttering to himself, and finally he turned and abruptly left the room. He returned ten minutes later and continued as if he had never left: "Yes, from this it obviously follows that...")

When he met Rutherford at one of the intercollegiate symposia/parties he immediately recognized that Rutherford's gruff,

3. Acceleration is a change in velocity, which involves both speed and direction. The electrons, while not necessarily changing their speed, are obviously changing direction and therefore accelerating.

jovial manner was the yin to his own yang, so he asked if he could come work with him. The rest, as they say, is history, for Rutherford's incomplete model of the nuclear atom quickly struck Bohr as the key to the behavior of electrons, and vice versa. "It could be that perhaps I have found out a little bit about the structure of atoms," he wrote to his mathematician brother, Harald. "If I'm right it would not be the indication of a possibility (like J. J. Thomson's theory) but perhaps a little bit of reality."

Bohr, like Einstein with his theory of relativity, was willing to focus his attention on a couple of points, discarding everything else, and then see where his reasoning led. In Bohr's case he began by accepting Rutherford's atom because nothing else could explain the helium-scattering experiments. He then turned to another set of data which no one else thought to connect with Rutherford's work: the line spectra of atoms.

When things get hot they give off light, as everyone knows. Solid objects shine over a continuous range of wavelengths, but when individual atoms are heated (as in a gas) their emitted light comes off in separate and discrete wavelengths. As measured on a photographic plate, it consists of a series of distinct lines. Each element has its own characteristic "line spectrum," comprising a veritable fingerprint of the emitting element; at the time we are speaking of this was well known, and in fact was used to identify the chemistry of unknown substances. Remember how helium was discovered in the sun by its characteristic spectral line? But no one understood why the emitted light should come off at separate wave lengths instead of in a continuous blurring.

Bohr started with two incontrovertible observations: the scattering of helium particles and the line spectra of atoms. He combined these with Einstein's new understanding of Planck's equation, which had been formulated to explain the continuous emission of light by black bodies.

The term "black body" is an imaginary but useful one; think "all men are created equal" or "an informed electorate." Like these, it is the ideal to be striven for; in this case it is the ideal

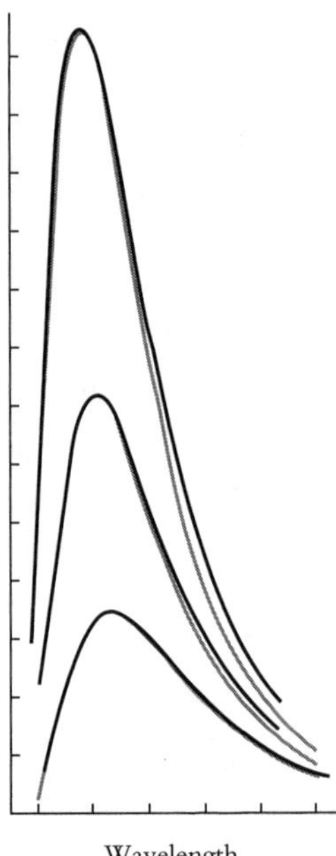

FIGURE 1. UV Catastrophe

emitter of light, with radiation determined solely by its temperature, and unlike those other examples can be closely approximated in real life. In all such experiments the emitted light forms a continuous curve when the wave length is plotted along the x-axis and the intensity on the y-axis, as shown in figure 1 for various temperatures.

An Austrian physicist, Wilhelm Wien, formulated an ad hoc equation that described these curves perfectly, but without any theoretical basis. The German physicist, Max Planck, dissatisfied with such a circumstance, set out to derive an explanation based on his interpretation of the second law of thermodynamics, which, in simple terms, explains the tendency of the universe toward chaos (increasing entropy). The prevailing explanation of this tendency was provided by Ludwig Boltzmann, with reasoning based on the atomic hypothesis; the second law resulted from the statistical behavior of large numbers of atoms. Planck regarded the universe as being ruled by something more precise than statistical behavior; indeed, in Boltzmann's description it was certainly possible (though unlikely) for the second law to be violated if sufficiently few atoms were involved. This was anathema to Planck, who wanted to show that the second law was inviolable.

He succeeded in his quest,[4] with one little flaw. In order to fit the experimental curves he had to assume that the radiation was emitted in small chunks, or quanta, rather than continuously. Without this assumption the curves at low (ultraviolet) wavelengths rose to infinity (a condition known as the "ultraviolet catastrophe") rather than dropping to zero, as the above graph shows that they do. Planck didn't look on these energy quanta as real, but solely as a needed mathematical contrivance.

Enter Einstein, who felt that such a contrivance ruined what was otherwise a beautiful theory. But if the quanta were real, the beauty was restored—and a whole new window into reality was opened: the quantum theory. His use of energy quantization to explain another experiment that had defied explanation (the photoelectric effect) earned him the Nobel Prize and established quantum theory as something to be dealt with.

Now Niels Bohr took up this quantum stuff and applied it to the two experiments. With the courage of a madman he suggested we forget everything else we think we know about atomic structure, and just account for these two experimental results. And it could be done simply, he realized, if we just postulate that the electrons circle the nucleus as Rutherford says, but their energies are quantized, as Einstein and Planck imply. This means that they can exist only in specified, discrete orbits. When they are heated they jump to an outer orbit and then fall back, emitting discrete bursts of radiation with energies corresponding to the difference in the energy levels of the orbits involved. And here comes the startling bit: when not being actively heated, the electrons stay in

4. A full description is beyond the scope of this book. See Helge Kragh, "Max Planck: The Reluctant Revolutionary," *Physics World*, December 2000.

their assigned orbits and do not emit radiation even though they are "accelerating." This assumption frees us from the collapse of the atoms.

But it contradicts Maxwell's theory, upsetting nearly everyone. How can an accelerating electric charge not emit radiation? Well, Bohr says, they just don't. He replied to his critics as Ring Lardner once wrote, in another situation:

> "Shut up," he explained.

Rutherford and a few others accepted this explanation, arguing that we were now entering a new realm of understanding and that further theories would be needed to understand what was happening inside the atom, where classical physics simply didn't apply. And shortly afterwards Erwin Schrödinger of Vienna supplied this theory: wave mechanics, in which the electrons were described as standing waves rather than particles. An electron in orbit was a wave "in phase" with the circumference of the orbit, that is, with a whole number of wavelengths exactly fitting the orbit. As such, it was perfectly in place, resonating with itself, and felt no need to radiate energy.

Light waves can be particles, or quanta, as Einstein showed, and particles (electrons) can be waves! Quantum theory was off and running, and hasn't stopped yet. And it all began with a "particle" of helium bouncing backwards off a metal foil.

SEVEN

Interlude: Helium, Argon, and Creationism

That old black magic
Has me in its spell...

—Johnny Mercer

HENRY M. MORRIS, WIDELY REGARDED AS THE founder of the modern creationist movement, died February 25, 2006, at the age of eighty-seven. His 1961 book *The Genesis Flood*, subtitled, *The Biblical Record and Its Scientific Implications*, was a cornerstone of the movement. Many more books followed, including *Scientific Creationism*; *What Is Creation Science?*; *Men of Science; Men of God*; *History of Modern Creationism*; *The Long War Against God*; and *Biblical Creationism*. In 1970 he founded the Institute for Creation Research, which continues to be a leading creationist force, now headed by his sons, John and Henry III.

In 1982 I debated the subject with him at the Coral Ridge Presbyterian Church in Fort Lauderdale in front of a sellout crowd of several thousand. He had emphasized in our initial contacts that the debate would be based on science, not religion, but when he opened his remarks with this same statement and the audience responded with loud cries of "Amen!" and "Praise Jesus!" I knew I was in for a long night.

Both of us steered away from the biological arguments, I because I'm not a biologist and he presumably because the Biblical side of that is so evidently silly—if he had tried to describe how Noah brought two mosquitoes or two fleas aboard he might have got away with it, but the whole panoply of billions of species of submicroscopic creatures was obviously a problem. Instead he concentrated on the physical side, in particular on the age of the earth, and that was fine with me.

As noted in the previous chapters, the earth's age is central to Darwin's argument. A strict interpretation of the Bible gives a limit of thousands of years, which is clearly not enough time for evolution to take place. Radioactive dating, on the other hand, gives Darwin his needed time span of billions of years, and so a cornerstone of the creationist argument is its necessary destruction.

Morris was a wonderful motivational speaker, and spent a long introduction wandering through the Bible to show how wonderfully reasonable it is. How could you look out at the universe—which is what scientists should do, he reminded us (as if they don't!)—and not understand that it couldn't have just happened? If you see an automobile, an airplane, a computer, could you imagine that they just accidentally happened, or is it not clear that someone created them? And how much more complex, more wonderful, are our universe, our world, our bodies and souls? ("Amen, Lord!" and "Praise Jesus!" rock the rafters.)

Then he got down to serious business. The scientists tell us, he told us, that the earth is billions of years old. (Anguished cries from the audience.) He held up his hands for silence, and begged for understanding. We are here, he said, to look at both sides fairly, and he walked across the stage and shook my hand as the audience cheered. "Now then," he addressed me personally, "you do claim that the earth is billions, rather than thousands, of years old?"

I agreed.

"Based on what you call radioactive dating? Such as the change from potassium atoms to argon?"

Yes, I agreed.

"In fact," he now turned his attention back to the audience, "the standard scientific claim is that argon, a gas which makes up one percent of the earth's atmosphere, has its origin in the radioactive decay of potassium within the earth, and when one calculates how much potassium is in the earth it is clear that it would take billions of years for that much argon to be formed." Then suddenly, as the prosecutor Sam Waterston might do on *Law and Order*, he swung back to me: "Isn't that true?"

His style was brilliant; when I agreed with him it seemed as if I was admitting a deficiency in my argument. And now, having set his trap, he sprung it: "But isn't it also true that potassium is not the only radioactive element in the earth? I'm sure you are aware of the element uranium, and that uranium in its radioactive disintegrations gives rise to another gas, helium!"

And now he turned in triumph to the audience. "I have calculated—(he swings back to me) I can show you these calculations right here and now—(back to the audience) I have calculated the production of both these gases, and it is perfectly clear that more helium is thus produced than argon! If the scientists are right, and the one percent of the atmosphere that is argon was formed from potassium over billions of years, then in that same time enough helium would be produced to form nearly ten percent of the atmosphere! And yet, (quietly now, with even perhaps a bit of sadness for the poor scientists) helium consists of less than one thousandth of one percent of the atmosphere."

(Loud clapping. "Praise Jesus!" etc.)

"If you would like to see my calculations," he said to me, "if you think I might have made a mistake . . . ?"

I shook my head and rose to my feet. The calculations were not in doubt, they were well known. "Dr. Morris is right," I said to the audience, and received my first applause of the night. When it died down I went on: "And Dr. Morris is wrong."

They didn't like that, but I plowed ahead. "Helium is produced in more abundance than argon, as every scientist knows. But that is not the heart of the problem. In fact, it is irrelevant. The real question is, how does the earth hold on to its gases?

"Well, of course, by gravity," I answered myself. "All the gases on earth would fly off into space if they weren't held by gravity. And our gravity depends on the mass of the earth. That is why the moon has no atmosphere: its gravity isn't strong enough to hold onto the molecules of gas.

"But it's also true that the effect of gravity depends not only on the earth's mass but on the mass of the molecules we're talking about. The heavier the molecule, the stronger the force of gravity holding it. And a very simple calculation [well, really not so simple, but what the hell] shows clearly that the mass of the earth exerts enough gravity to hold our heavier gases but not our lighter ones. The breaking point is roughly at the mass of the water molecule; anything much lighter than this is too light to be held by the earth's gravity. That is why there are so few hydrogen molecules on earth, though hydrogen is by far the most abundant element in the universe. Unless it's tied up and bound in heavier molecules, the hydrogen just drifts off the earth and out into space. And next to hydrogen, the lightest molecule is helium. Its average lifetime in our atmosphere is on the order of millions of years, and so over the billion-year lifetime of the earth nearly all of it has been lost to space. Argon, on the other hand, is twice as heavy as water, and so it's retained and builds up over time."

I couldn't believe Morris hadn't known that. I now believe that he did, but he also knew that his audience didn't. Still, he would have known that I would know that and so his argument would be easily destroyed. I thought he must not be as smart as I had anticipated, and that this would be an easy evening after all. How wrong I was!

He paid no attention to my refutation but simply went charging on, and as he knew it would, the audience forgot my argument and followed his: "I believe you said that the argon builds up over time? By time you mean billions of years? And how do you know that?"

Before I could begin to answer he went on: "By this method of radioactive decay! This is the cornerstone of your entire argument! Is this not true?"

Yes, I admitted.

He whirled around and dove into the pile of books and papers he had brought with him, and came out with a yellowish journal held over his head. "I have here a copy of the *Journal of Geophysical Research*," he intoned. Turning to me, he asked, "Is this a respected scientific journal?"

"Of course."

"It does not publish idle speculation, or even religion?" (He turns to the audience; laughter. Oh, we're having a good time.) "It publishes good, solid science, am I right? Science, based on evidence, not belief!"

When I agreed, he opened the page and read this sentence: "We conclude that the high ages, based on potassium argon dating, are not valid but are probably due to some unknown mechanism introducing excess argon into the system." He turned to me and recited by heart, emphasizing each word, "The high ages, based on potassium argon dating..." and here he turned back to the audience, triumphantly: "are not valid!"

The audience audibly gasped, whispered to themselves. He waited till they quieted, then said, rather sadly, gentlemanly apologetic in his triumph, "Science has spoken." And he turned and sat down.

But something about that sentence was familiar. I thought I recognized it. "What's the title of that article?" I asked, and he read it off, "The potassium argon problem in iron meteorites," emphasizing the word *problem*.

I didn't want to compare myself to Huxley, but all I could think of was his phrase: "The Lord has delivered him into my hands." Praise Jesus!

"Who wrote that article?" I asked.

He hadn't noticed. He looked at it now. "Rancitelli and Fisher," he said, "from the University of..." he paused... "Miami."

Aha, he noticed.

"Yes," I said, "Fisher, that's me. It's my article. Do you really think I was saying that potassium argon dating is wrong? Quite the opposite." And I explained to the audience that you can't simply pick up anything at all and date it; you have to know a lot about it first. The technique is most suitable for volcanic rocks, which exude their accumulated argon upon eruption, thus setting their clock to zero. Argon formed from decay of potassium after that event measures the time of eruption, and measurements on terrestrial volcanics give reliable ages up to billions of years. What we were trying to do in the quoted paper was test whether the method was applicable to iron meteorites, and we decided that it wasn't.

"That's the strength of the scientific method," I explained. "You test things, so you know when a given method is reliable or not. We don't just measure potassium and argon and have blind faith in the numbers, we test them! And when a method doesn't work, we say so. Iron meteorites are difficult to date, but stone meteorites—and our earth—are not. The methods have been tested and they give reliable dates: our world was created four and a half billion years ago."

When it was over I felt I had completely destroyed his "scientific" arguments and showed that his case rested entirely on religion. You either believed that the Bible was the totally accurate word of God and that all science was either wrong or irrelevant, or you didn't. But afterwards, in the car driving home, my wife told me another story. She had been in the audience, and she listened to what they were saying as they left the church. The tide was overwhelmingly in Morris's favor. The one remark that came closest to understanding the science was when one mother chuckled to her family, "Dr. Morris really showed that scientist what's what, didn't he?" Her husband said, "You know, I'm not sure. Morris didn't really answer—" And then his wife glowered, and he quickly shut up. His daughter piped up with "Daddy, you can't argue against the Bible, can you?" And he quickly replied that of course you can't.

Everyone knows that.

꧁

Postscript: Several years later I heard Dr. Morris being interviewed on the radio, and he presented the same arguments that I had refuted: he talked about argon and helium in the atmosphere, and he even read that same sentence from my paper about the deficiency in potassium-argon dating. He knew that those arguments were wrong, but he also knew that most of his audience would not know that. He flat-out lied to them.

What was it Barry Goldwater was saying back in 1966? "Extremism in the defense of liberty is no vice!" I guess that applies also to lying in the service of the Lord.

EIGHT

Meanwhile, Back at Brookhaven

Im Westen Nichts Neues (All Quiet on the Western Front)
—Erich Maria Remarque

AT THE END OF THE NINETEENTH CENTURY WILLIAM Ramsay, searching for minerals that might concentrate argon or helium, wrote, "One mineral—malacone—gave appreciable quantities of argon; and it is noteworthy that argon was not found except in it (and, curiously, in much larger amount than helium), and in a specimen of meteoric iron. Other specimens of meteoric iron were examined, but were found to contain mainly hydrogen, with no trace of either argon or helium. It is probable that the sources of meteorites might be traced in this manner, and that each could be relegated to its particular swarm."

Finally, sixty years later, this is what Ollie Schaeffer and I now set out to do.

Meteoritic iron has been used since prehistoric times: necklaces of the metal beads interlaced with gold are found in the tombs of Egyptian kings, and an inventory of a Hittite temple, describing where on earth their gold and silver came from, lists their iron as having "fallen from the sky." Yet as late as the early nineteenth century, the reality of meteorites still was not accepted by men of good will. For after all, how could heavy stones and chunks of iron fall out of the sky? And then in 1803 a huge shower of meteorites fell at L'Aigle, France, just at the time that the French Academy of Sciences had convened a meeting to discuss the question.

In America no one paid much attention, until on December 14, 1807, at 6:30 in the morning, a bright fireball suddenly blazed through the sky over Vermont and Massachusetts. It was reportedly nearly as bright and big as the moon, until it suddenly exploded and disappeared over the town of Weston, Connecticut, showering the area with stone fragments, as the local media reported.[1]

In those days it took a while for the news to travel a few tens of miles, and so it was a few days before Yale's new professor of chymistry (*sic*) and natural history, Benjamin Silliman, heard of it. Grabbing his hat and a colleague, Professor James L. Kingsley, he galloped across the state to Weston. Together they talked to everyone who claimed to have seen the phenomenon, and found that several stones had been recovered. Although American men of science still weren't sure that meteorites existed, the local populace had no doubts. They had heard the ancient stories, had (as people tend to do) gotten them a bit twisted, and had greedily smashed the stones to retrieve their hidden riches: "Strongly impressed with the idea that these stones contained gold and silver, they subjected them to all the tortures of ancient alchemy, and the goldsmith's crucible, the forge, and the blacksmith's anvil, were employed in vain to elicit riches which existed only in the imagination." Nevertheless the two professors managed to bring back to Yale "a considerable number" of specimens.

1. At a meeting of the Meteoritical Society a Swiss scientist gave a talk about his noble gas measurements on the Weston meteorite, and began by saying, "*Im Weston nichts neues,*" which was about the funniest thing I had ever heard at a scientific meeting. Nobody else laughed; sic transit gloria mundi.

Silliman analyzed the meteorite chemically and published the results in the *Transactions* of the American Philosophical Society in 1809. "The case was deemed so interesting and important that the published account was read aloud in the Philosophical Society of London & in the Academy of Sciences of Paris. It was admitted to be one of the most extensive and best attested occurrences of the kind that has happened and of which a record has been preserved."

Not everyone was convinced. Down in Virginia, Thomas Jefferson, being apprised of their work, famously (and possibly apocryphally) replied that he "found it easier to believe that two Yankee professors might lie than that stones might fall out of the sky." But by the end of the century Jefferson was dead and the existence of meteorites was generally accepted, though only in general terms. They did indeed fall out of the sky, but from where did they come? Thus Ramsay's suggestion that argon and helium might help settle the problem.

But not much happened until in 1933 Hitler came to power in Germany, and among the scientists who fled to England was Friedrich Paneth. In 1942, in the midst of the Second World War, he began the first study of the noble gases in meteorites by trying to apply Ramsay's ideas on radioactive dating to the iron meteorites. At that time we had no definitive proof that the irons were an integral part of the solar system, and some people thought they might be intruders from deep space. Paneth's idea was that if they showed the same age as the earth (Rutherford's estimate of about three billion years), it would indicate that they had formed at the same time and presumably in the same process, while if they were older they must have originated differently and presumably elsewhere. If they showed younger ages it would prove nothing, since there was always the possibility of helium loss through diffusion.

Well, two out of three chances of learning something is definitely worthwhile, so he got to work. The result was that the uranium-helium ages[2] of several irons ranged from a few million years to greater than six billion—much older than any estimate of the earth's age, leading to the supposition that different parts of the solar system had different origins. The war then intervened, and the problem of the origin of the solar system was put on hold while Paneth worked on another problem quite as interesting: he joined the British research team working on the creation of the atomic bomb. Not until after the war did he return to Durham and try to put his lab back in working order. While he was doing so, Carl Bauer in the United States and Harold Huntley in England independently identified the problem: they suggested that some of the helium might be of cosmic ray origin—that is, formed not by the decay of uranium but by nuclear reactions induced by high-energy cosmic rays—and that therefore the uranium/helium ages were invalid.[3] But how to separate the helium arising in the radioactive decay of uranium (radiogenic) from that produced by cosmic rays (cosmogenic)?

Not to worry. Bauer pointed out that radiogenic helium consists almost entirely of helium of mass 4 (^4He), while the cosmic rays should also produce the lighter isotope of mass 3 (^3He). Then Fred Singer,[4] a brilliant young physicist on duty at the time with the Naval Research Office liaising with the United States Embassy in London, calculated the relative production rates of the

2. He tried to measure potassium-argon ages, but there was too much contamination from atmospheric argon, always a problem.

3. As it later turned out, the uranium measurements were also in error.

4. He is one of the smartest men I know. He would later predict the existence of the Van Allen belt (some of us think it should be called the Singer belt), but what really impressed me when I met him years later was his attitude toward the *Journal of Geophysical Research*. This journal came out twice a month (in those days; more frequently now) and carried a large number of papers ranging from theoretical studies on the atmosphere to geochemical work on the ocean bottom and everything in between, and I was lamenting that it was a waste of paper to send it to all of us because no one could read more than a few of the articles in any issue. He looked at me askance and said he didn't know what I was talking about. He himself read every article. (He is the only intelligent person I know who isn't worried about global warming. I hope fervently that he's right, but I doubt it.)

two isotopes, arriving at values of about 0.3 for the $^3He/^4He$ ratio. Paneth immediately confirmed this by measuring the ratio in several meteorites and finding results clustering around the predicted value. Bauer and Singer also predicted that as the cosmic rays penetrated the iron mass they would lose energy, changing the specific effects of the nuclear reactions they induced. Ollie Schaeffer and I thought this would result in changing helium/argon ratios, which I thought would enable us to study the intricacies of nuclear reactions induced by high-energy protons on iron by substituting the iron meteorites and the cosmic rays for manmade targets and the nonworking Cosmotron.

Alas, it was not to be. The uncertainties were too many and too overwhelming: we didn't know the flux of cosmic rays or how it varied in time and space, we didn't know the nuclear production rates over the cosmic energy spectrum, and we didn't know how the cosmic rays varied in intensity and energy as they penetrated into the depths of the meteorite. We didn't even know what the original depths of the meteoritic samples were. Instead we took up Ollie's suggestion (following Ramsay's original idea) that we might be able to use what we knew about nuclear reactions to study the origin and history of the meteorites.

The problem Ollie and I ran into was related to the origin of the meteorites. At the time it was supposed that the irons might come from the asteroid belt and the stones from the surface of the moon, or perhaps they both came from comets. The asteroid-belt model visualized the asteroids as the remnants of a disrupted planet, with the iron meteorites representing the core of that planet. As the remnants whirled around and crashed into each other, some might get knocked out of their orbit and fall into the gravitational field of the earth, ultimately winding up in a crash onto our surface as iron meteorites. Others might smash into the moon, knocking out chunks that would then fall to earth as stone meteorites.

Or perhaps the stone meteorites were also broken asteroids. Or perhaps they both came from comets (nobody had any idea what comets were made of or how they originated.)

The field was wide open, but one thing was clear: the meteorites were broken parts of something, and the important words there are "broken parts." The cosmic rays, as far as nuclear reactions are concerned, are Mev- to Gev-range protons, and as these pass through a mass of material, some of them hit a nucleus and break it apart while others bounce off the atoms, slow down, and eventually smash another nucleus but at a much lesser energy. In addition, the first nuclear reaction products would be knocked out at high enough energies to then induce other reactions which would in turn induce other reactions...

> So, Nat'ralists observe, a Flea
> Hath smaller Fleas that on him prey,
> And these have smaller yet to bite 'em,
> And so proceed *ad infinitum*...[5]

In fact, the nuclear reactions in the meteorites would occur at a continuous range of energies, and how could we pick them apart and make sense of them with regard to nuclear theory if we didn't know their original size, or whether they had broken to their present size in just one collision at some particular time or were gradually worn down in a series of many collisions over millions or even billions of years?[6]

As it turned out, we couldn't.

Instead, what Ollie suggested as I sat there shaking my head in misery was that we could turn the problem around and instead of using the meteorites to study nuclear theory we could use what we knew of nuclear theory to study the history of the meteorites. As Ramsay had said, we might be able to "trace the sources of

5. This is not by Ogden Nash.

6. A further complication, if one were needed, is that the meteorites melt away some unknown portion of their surface as they burn their way through the earth's atmosphere.

meteorites in this manner, [so] that each could be relegated to its particular swarm." Or as Singer had put it, "It is hoped that further measurements of helium isotopic ratios in meteorite samples can establish systematic groupings not only of meteorite ages but also of times of breakup. Such indications would be of the greatest importance for theories of the origin of meteorites."

Well, it was worth a try.

In order to study the history of nuclear reactions in meteorites, the noble gases were the only way to go, because the amount of material produced by such reactions is vanishingly small; only the noble gases were low enough in natural abundance—remember, that's why they're also called "the rare gases"—so that the small amounts of cosmogenic isotopes can be accurately assessed.[7] The choice of meteorites to study was also a no-brainer. In the study of nuclear reactions the target element is a basic consideration, and the irons are composed almost entirely of iron and nickel, while the stones are a complex mélange of calcium, aluminum, silicon, oxygen, etc. So we started our work with the irons. The first one we looked at was a strange beast called the Washington County meteorite, strange because not everyone thought it was a meteorite. It looked more like a piece of slag metal, but when we found a large amount of ^{3}He in it we knew conclusively that it was a meteorite. (This was reported on the front page of the *New York Times*, and I thought that science was really going to be fun; as things turned out, it was the last time any of my research made the papers. On the other hand, science really has been fun.)

By this time it had been established that all meteorites, both stones and irons, had formed at the same time as the earth, some

7. One group in Germany was beginning to use potassium isotopes.

four and a half billion years ago. This is known as their solidification age. Since a size on the order of tens of meters would be enough to shield the interior from cosmic rays, it was possible to define a "cosmic-ray age" as the time since the meteorites were broken into fragments small enough to be exposed to the cosmic radiation. The first very rough determinations showed that the cosmic ray ages of all meteorites were much lower than the solidification age, which meant that the meteorites hadn't been created at their present sizes; they were fragments of larger pieces that had been broken apart relatively recently in solar system history (millions of years ago, rather than billions). In addition, in contrast to the single solidification age they showed a variety of ages, with stones apparently younger than irons and with only one iron reaching the billion-year range.

So this was what we wanted to get a handle on, and Ollie came up with a neat idea. Measuring the noble gases that were spallation products, the products of the cosmic ray–induced nuclear reactions, wasn't tough, but in order to calculate an age we had to know their production rate. This could be calculated from the cosmic ray flux and energy spectrum if we knew them, but we didn't; they were unknown quantities. But it had been suggested previously that if a radioactive isotope is being produced constantly for a time long compared to its half-life, its rate of disintegration will constantly rise until it reaches the production rate and will then continue at that rate. In other words, its formation and disintegration are in equilibrium and will continue to be so as long as the production continues at the same rate.

So if the cosmic ray flux is constant (which seemed reasonable, even if the figure was unknown), a radioactive isotope in the meteorite would give its production rate and then measurement of a stable isotope would give its total production, and a simple ratio of the two would give the cosmic ray age T:

$$T = \text{Total}/\text{rate}$$

But only if the two isotopes have the same production rate. The first attempts were made using the pair $^{3}He/^{3}H$, where ^{3}H (tritium) is a radioactive isotope of hydrogen with a half-life of twelve years. Being so close together in mass, the production rates should be very similar. But we were worried about some other problems intrinsic to the method: some of the cosmic rays are composed of helium, upsetting the cosmogenic ratio; small amounts of ^{6}Li in the meteorite would upset the balance by permitting other helium-producing reactions; and perhaps most important of all, both ^{3}He and ^{3}H are small atoms and could possibly leak out of the meteorite—and no one knew which might leak more easily, so no one could predict the result.

Ollie suggested using another stable/radioactive pair, ^{36}Cl and ^{36}Ar. Esther Sprenkel, also on a postdoctoral in the department (she was known as Big Esther in contrast to Little Esther, another postdoc), was measuring the radioactive isotope ^{36}Cl in meteorites; together with our measurements on ^{36}Ar, this would obviate all the helium uncertainties. And indeed it did, but the results were less than world-shattering.

The iron meteorites showed a spread in ages, indicating that they broke into fragments continually over time and much later in solar system history than the original formation date of 4.5 billion years. But, ho hum, this gave little new information on meteoritic or solar system history; the results were "not inconsistent" with what we knew before: they could be fragmented asteroids or cometary nuclei broken free, or whatever.

In the meantime, some other workers were silly enough to try to measure cosmic ray ages on stone meteorites. They had to rely on the $^{3}He/^{3}H$ pair, because ^{36}Ar and ^{36}Cl are produced only from components of higher mass; for meteorites this means iron and nickel, and there weren't enough of these elements in stones to be accurately measured by the techniques available at the time. But as it turned out, all our worries were for naught and they got good ages for the stones, which, in contrast to the irons, clustered around a few specific values, indicating specific events in solar system history. Particular groups of stone meteorites have evidently formed in specific breakups. It was still not clear whether the parent bodies were asteroids or cometary nuclei, but current research by unmanned flybys of the asteroids indicate that they are chemically identical to some of the stone meteorites, so at least some of the meteorites had their origin there, and other (non-noble) gas work shows that the asteroids are not remnants of a disrupted planet but rather material that never formed into a planet (probably because of the disruptive effect of Jupiter's gravity).

The noble gas data clearly showed that the spectrum of iron meteorite ages is distinctly different from that of the stones: not only is there no clustering, but the ages are older than the stones. The reason for this was a subject of controversy for several decades. The obvious answer was that the irons are simply older, meaning that they had broken into meteoritic size hundreds of millions of years sooner than the stones, which would mean they had different origins in space as well as time; perhaps the irons were cometary and the stones asteroidal.

But other explanations were possible, and rather more likely. Eventually I was able to modify a proposal called space erosion to explain the difference. Originally this was an idea put forth by Ed Fireman and Fred Whipple at the Smithsonian Astrophysical Observatory to explain the difference in cosmic ray ages of irons as worked out by two different methods. The concept was that the irons would be eroded during their hundreds of millions of years in space by on the order of 10^{-8} cm/yr, and this would affect different methods of calculating ages in different proportions. It was shown to be wrong, but I later showed that stone meteorites were brittle enough to be eroded—or destroyed—completely within a few tens of millions of years, and this effectively accounted for the difference between them and the irons: they simply didn't last long enough to reach the ages of the irons. A group at Berne came up with a similar idea almost simultaneously and expressed it better as "collisional destruction" instead of erosion, with more of an emphasis on occasional collisions with larger objects rather than continuous erosion by small particles, and so the idea has been accepted.

With that, I began to think there was nothing much more of interest to be garnered from the study of cosmic rays in meteorites, and I was getting sick and tired of the mass spectrometer, which, despite its early promise of lots of data with little effort, was proving a recalcitrant monster. The glass line was prone to infinitesimal leaks that took days or weeks to find and fix. The mercury pumps spilled droplets on the floor, which took hands-and-knees technology to clean up. Our continued efforts to measure production rates in the Cosmotron for comparison with the meteoritic data led to sleepless nights and despondent days. As my two-year appointment at Brookhaven neared its end, I was too often spending a whole day tuning up the mass spec and then turning one knob a touch too little or a dab too much and wham!—I'd crack the glass and be a week behind where I had been when I woke up that morning. So when my postdoctoral at Brookhaven expired in 1960 and I joined the faculty of Cornell as an assistant professor in physics, engineering physics (now applied physics) and the Center for Radiophysics and Space Research, I fondly and without regret bid adieu to the noble gases.

Cornell didn't have a noble gas mass spectrometer. It had instead a nuclear reactor, and I intended to embark on a new line of research: activation analysis. And the gods laughed.

NINE

Cornell, the Ten-Minute Experiment, and Back to Argon

Other arms reach out to me
Other eyes smile tenderly . . .

—Hoagy Carmichael

BECAUSE, YOU KNOW HOW IT IS, YOU NEVER FORGET your first love.

I tried, believe me, I tried. I turned enthusiastically to activation analysis, which is a technique in which you irradiate a sample in a nuclear reactor—which, unlike the Cosmotron, just sits there day and night pouring out neutrons which slip into the sample, hardly disturbing it but pushing some of the atoms over the line from stable to radioactive. Since radioactivity is the most sensitive detection type of them all, you can measure the presence of extremely small amounts of many elements in the most complex of minerals. Roman Schmitt of General Atomic (later at Oregon State University) and I had discussed the possibility of detecting some elements without any work. You would just drop the sample in the reactor and then stick it in one of those newfangled multichannel gamma detectors. There would be a whole mélange of gamma rays coming out from all the radioactive isotopes in the sample, but a few elements throw out gammas of very high energy, and these would stand out above the background noise, or so we thought. The data would give immediate results while we just sat there reading a book or daydreaming.

For instance, aluminum. It consists of a single isotope, ^{27}Al, and upon bathing it in a sea of neutrons it forms ^{28}Al, which is radioactive with a half-life of 2.3 minutes and a gamma ray of 1.78 Mev. Beautiful! Hardly any element throws out a gamma of energy more than 1 Mev, so the 1.78 bugger would stand out easily, and the 2.3-minute half-life meant that I could easily follow its decay for positive identification. Several half-lives, or, say, ten minutes, would be plenty. By irradiating a sample of known aluminum content, a comparison of the two activities would give the aluminum in the unknown.

What every scientist dreams about: the ten-minute experiment! The problem was to find some reason to measure aluminum. Who cares what the aluminum content of rocks is? The answer came as I was reading a wonderful paper by Harold Urey and Hans Suess (whom I mistakenly thought was the famous Dr. Seuss; he looked just like I expected the *Green Eggs and Ham* man to look) on the origin of the elements. Their thesis was that the chondritic[1] meteorites were the basic building blocks of the planets; for example, the chondritic-earth model, a favorite among geologists, envisages the earth as having been put together by the aggregation of chondritic stuff. Their evidence for this was that all the nonvolatile elements have identical abundances in all the chondrites, and this matches their abundance in the sun. (The volatile elements, such as the noble gases, were lost when the meteorites and the earth were hot, before they solidified, which is reasonable.)

1. Stone meteorites can be divided into two classes: chondrites, which contain embedded minerals called chondrules, and achondrites, which do not.

Sounds good. But then, being trained not to believe anyone,[2] I looked at the data they claimed as evidence. And lo and behold! The aluminum data didn't back them up. It took a month of searching through all kinds of records, but where Urey and Suess claimed that the aluminum content of chondrites was constant at 1%, I found a variety of abundances in the chemical literature, ranging from "undetectable" up to 15%. I wrote to Urey and asked him about this, and his answer was that those values were "unreliable." Why were they unreliable? Because they didn't fit his theory!

This is no way to do science. On the other hand,[3] aluminum is notoriously difficult to measure by classical chemistry. It forms a white gooey flocculate instead of a nice powdery precipitate, and it's hard to separate it, dry it, and weigh it. So maybe Urey is right and some of the published values are wrong, and maybe Urey's wrong and aluminum is variable in the meteorites, which would destroy his whole theory, because aluminum is a perfect example of a nonvolatile element.

The first thing to do was to test my new technique. I already had several meteorites which Urey had listed as "normal"; that is, with an aluminum content of 1%. I was so excited that, although it was Saturday, I ran over to the reactor lab to try it out. A grad student had the reactor scheduled that day, but I explained to him about this exciting possibility, and that I could measure all the meteorites and prove Suess and Urey were wrong and change all our ideas about the creation of the earth in one day, so he stepped aside and let me go first.[4] (He was happy to have an excuse to leave; after all, it was Saturday.) And it worked perfectly; by the

2. The Buddha advises "Do not believe in anything merely on the authority of your teachers and elders," which is good advice even if you're not a Buddhist.

3. There is always another hand to be reckoned with in science and politics. Harry Truman cried out in desperation for a one-handed general to give him advice with certainty, but in science that other hand makes for all the fun. (And what Truman got was MacArthur, so be careful what you wish for.)

4. This would have unexpected consequences. See chapter 10.

end of the day I had measured a half dozen meteorites and all gave the "correct" value of 1%, and I left, happy as a lark. Who needs a mass spectrometer?

But of course I had exaggerated about that "one day" business. The next step was to write to museums all over the world, to get as many of the "unreliable" meteorites as I could. This took a few months, but eventually I had a representative sampling. I took them in to the reactor one afternoon, and guess what? By evening I knew that Urey was right: every last one of them—with published analyses ranging from zero to 15%—gave an aluminum value of 1%.[5] On the nose.

I spent the next few years working with a few graduate students, doing activation analysis of several elements in meteorites with this nondestructive technique, as I called it.[6] But as time passed I felt the old pangs insistently returning... Like Hoagy Carmichael felt about Georgia, I had the noble gases on my mind.

After a year's solid research at Oak Ridge and two more at Brookhaven, I had decided that teaching might not be so bad. Originally I had thought that giving full time to research was the way to go, but when I realized that nobody at a national lab had ever won a Nobel Prize—they nearly all went to university professors—I began to think that maybe teaching wouldn't be a total waste of time after all.

So when my postdoc at Brookhaven expired I wasn't too upset when Oak Ridge turned me down. When I finished my thesis year there, I had been offered a permanent job, but they suggested

5. To be accurate, 1.3 +\− 0.1%.

6. The name came from the fact that the sample wasn't dissolved or melted, as it would be in all other methods of chemical analysis; it was just irradiated, and then the induced radioactivity was counted. Roman Schmitt named it Instrumental Activation Analysis, or INAA, and that's the name that has stuck.

I go elsewhere for a year or two to get more experience and to find out how other labs did things in different (and perhaps better) ways, and there'd always be a job there waiting for me. Which was a good idea, but when I called them now I was told sorry, this year the budget was tight and there weren't any openings. As Yogi Berra says, a verbal promise isn't worth the paper it's written on.

I had two job offers, one at Los Alamos National Lab and one at Cornell, and I chose the latter. I turned up in the chairman's office for my debut at ten o'clock one day in mid-September. He looked at his watch and remarked, "I'm glad you're here. Your class starts in half an hour." We hadn't even discussed what I would be teaching. It turned out to be a graduate course in Modern Physics, a course I had never taken myself. This set a precedent: as I look back, I realize I've never taught a course I've had as a student.

On our first Saturday night in Ithaca, my wife and I were standing in line at the local movie house, among a gaggle of students. The two in front of us were discussing their physics courses. One said his was the dumbest, most boring course he had ever had, the other said his own course was the most exciting he'd ever had. After a while they discovered they were both in the same course, but in different sections. The bored student had a new assistant professor for instructor, the excited student had Philip Morrison.[7] I decided then and there that maybe teaching was not just good training for research but might be worthwhile in itself. Thanks, Phil.

My salary had jumped from six thousand dollars to seventy-five hundred when I joined Cornell, but my wife was a little pissed because I had turned down a few offers from commercial firms of over ten thousand. (Those are yearly salaries, not monthly.) So

7. Phil Morrison was an outstanding physicist and exciting lecturer who died in 2005. See chapter 19.

I was happy to discover that it was possible to add another third to my income if I could get a research grant to cover my summer salary. The problem with that is you have to tell the agency what you're going to do, which means making plans, which I had never done.

But it turned out, in those days, to be not too hard. In my first months in Ithaca, I hadn't yet worked out the details of instrumental activation analysis, so I wasn't able to write a coherent proposal. I turned instead to some unfinished work from Brookhaven.

Our first meteorite there had been Washington County, and we had found a large amount of ^{4}He in it, too much to be accounted for by cosmic ray production. We had suggested that it came instead from the decay of uranium, although we noted that it could possibly be primordial helium left over from the original coalescence of the meteorite. But this seemed to me too exotic an explanation; I was sure it was simple decay of uranium, leading to the possibility of getting a uranium-helium age for the meteorite, which would be quite important because there were no valid solidification ages for iron meteorites.

The uranium content of iron meteorites is so low that the few older attempts to measure it were invalid. Terrestrial contamination is the problem, a problem that is virtually omnipresent in geochemical and cosmochemical work and was only then beginning to be appreciated. It arises because everything is dirty: there's a little (or more than a little) of everything in everything. You can pick up a piece of the purest metal that man knows how to make, and if your analytical instruments are sensitive enough they will find every single element in the universe in that piece of "pure" metal. And much more of everything is to be found in natural specimens of rocks or meteorites.

Previous to activation analysis and mass spectrometry, the "normal" methods of geochemical analysis could measure stuff down to roughly parts per million; that is, a millionth of a gram of, say, uranium, could be measured in a gram of rock. But suppose you had a meteorite with much less than that amount. Would you measure its uranium content as zero? No, you would not, because while that meteorite was sitting around in the dirt, before it was discovered, uranium from the surrounding dirt would find little pathways into its interior, and when you measured it you would find parts per million in it—but all you would be measuring would be the terrestrial contamination.

This is a constant and sometimes overwhelming problem, which was only beginning to be recognized. During my time at Cornell, a professor at Fordham found convincing evidence of fossilized life in a meteorite from the Museum of Natural History in New York. The "life forms" turned out to be, after much work and argument, terrestrial pollen grains which had infiltrated the meteorite from the New York air.

Activation analysis is capable of measuring stuff down to the sub-ppb (parts per billion) level, and so contamination always has to be considered. With careful preparation, one can usually—but not always!—avoid it. At any rate, it's the preferred technique for elements expected to be very low in abundance.

The only activation analysis attempt at measuring uranium in iron meteorites had looked at just a few irons and found contents so low that the older methods couldn't possibly have captured them; those older results must have been due to the contamination that is so easy to pick up both during museum storage and even during the chemical analysis. So I carried out a few basic calculations and showed that if the Washington County meteorite had solidified four and a half billion years ago, as it almost assuredly had, then the amount of uranium necessary to produce the observed helium was so low as to be measurable only by activation analysis. The Atomic Energy Commission (AEC) immediately gave me the money, throwing in some travel funds in case I wanted to do the work elsewhere.

I certainly did. Cornell was fine, but my wife was homesick for Hartford. Ithaca was a boring town at best, and I hadn't yet set up my own lab. Ed Fireman (see "space erosion" above) had been a postdoc at Brookhaven just a year or two ahead of me and was now at the Smithsonian Astrophysical Observatory. I told him what I was planning to do, and he suggested I come to Cambridge, use the MIT reactor, and do the work in his lab. And so I moved the family (our daughter Liz had been born at Brookhaven and our first son Ronald had been born just the previous winter) to Hartford, and I left them for four days a week to work at the SAO (now the CfA, the Harvard-Smithsonian Center for Astrophysics).

The work would be a noble gas study, but substituting activation analysis for the mass spec, which wasn't capable of the low levels of measurement we needed. We would irradiate the sample in the reactor, causing the uranium to fission. One of the fission products is a radioactive isotope of xenon, which we'd measure in a specially fabricated Geiger counter that we'd attach directly to the extraction line.[8] The first thing to do was get a sample of Washington County, which turned out to be impossible. We had used up all we had at Brookhaven, and there wasn't any to be spared at any museum. But by the time we had run out of museums to ask, we had completed the vacuum extraction line and had built and tested the Geiger counters, and I had to have something to tell the AEC about how I had spent their money, so Fireman suggested we measure another meteorite to test a suggestion recently made by some Russian scientists.

The earth and the stone meteorites were well dated by lead isotopes at 4.5 billion years—but the irons only by inference. The *assumption* was that they had formed at the same time and from the same isotopic mix as the stone meteorites and the earth, an assumption based on lead-lead ages. Two isotopes of uranium, ^{235}U and ^{238}U, decay through a series of steps to the lead isotopes ^{207}Pb and ^{206}Pb, respectively. Since the uranium isotopes decay with different half-lives, the radiogenic lead isotopes grow into the meteorites at different rates, which means that the age—the lead-lead

8. After irradiation we'd put the sample in a vacuum line, melt it, clean up all the gases except the noble gases, separate xenon from the others, and pump it into the counter.

age—can be determined by measuring the ratios of the lead isotopes without measuring the uranium at all, and this could be done very precisely. The only glitch is that you have to know what the original, or primordial, lead isotopic ratio was.

Measurements of the stone meteorites and terrestrial lead gave data that led to a straight line (figure 2).

The iron meteorites lie on the same line, with the lowest amounts of ^{206}Pb and ^{207}Pb. Since the (admittedly sparse) uranium data in the irons indicated insufficient uranium to account for their amounts in any reasonable time, the assumption was made that they represent the original ratio when the solar system formed. With this assumption, the data make a logical and self-consistent story: all the solid bodies of the solar system began with the lead found in iron meteorites, and four and a half billion years later the earth and the stones have evolved to their present composition while the irons have remained primitive because they don't have enough uranium to change them in any measurable way. In this scenario, all meteorites have the same age: 4.5 billion years.

But I. E. Starik and his colleagues in Russia found radiogenic lead in a few iron meteorites. The composition of their lead was similar to that on earth, giving them the same age (4.5×10^9 years),

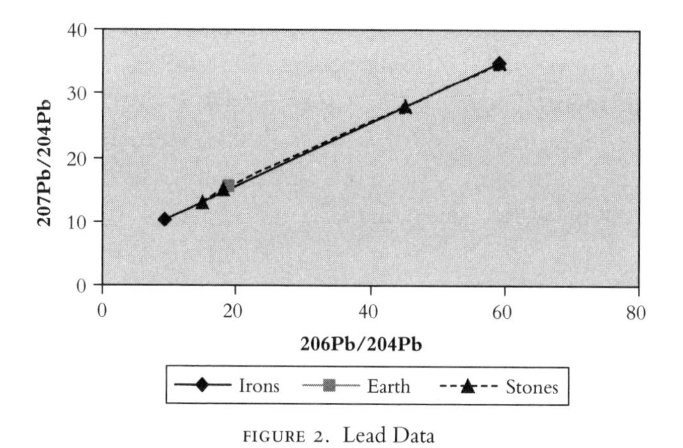

FIGURE 2. Lead Data

but only if they had enough uranium in them to have produced the lead. And the only uranium data on iron meteorites was too low to account for the growth in lead over that time. *On the other hand,* none of the uranium measurements had been made on any of Starik's irons that had the radiogenic lead. Luckily we had one of them, a large meteorite that had smashed into a region of southeast Siberia called Sikhote-Alin and had broken into a million or so chunks, pieces of which had been widely distributed.

By this time there were only a few weeks left before summer was over and I'd have to be back at Cornell, but the experiment went nicely, and we found results similar to those the Chicago group had measured on other irons: the uranium content of Sikhote-Alin was less than 10^{-11} gm/gm, a hundredth of a part per billion, orders of magnitude too low to have provided the radiogenic lead in less than ten billion years, an unacceptably high age. Starik's group had suggested that there were two groups of iron meteorites with two different primordial isotopic compositions, which contradicted the common assumption made in calculating meteorite ages, that the iron and stone meteorites and the earth all have the same primordial isotopic composition of lead, indicating that the whole solar system formed from the same material at the same time, and we reluctantly concluded that this couldn't be ruled out by the data, although it certainly would complicate our ideas of solar system formation if their data were valid. More reasonably, their data could be explained simply by terrestrial lead contamination. But what if...

Unreasonably, my thoughts as I drove back to Cornell with my wife, a sleeping baby boy, and a bright-eyed two-year-old daughter who read every sign as we passed, driving me crazy, were... well, my thoughts were crazier still.

My thoughts were slowly, reluctantly, excitedly focusing on something spectacular. I knew it was crazy, but—

The amount of uranium we had measured in Sikhote-Alin was enough to produce the lead in ten billion years, which was much too great an age to be real. Except, I was thinking...

If the uranium content of Washington County was the same as Sikhote-Alin, its uranium-helium age would be—ten billion years! Again, much too great to be real. But I couldn't help thinking...

My predecessor in Ollie Schaeffer's lab had been a German postdoc, Josef Zähringer, who had hooked up with the department's analytical chemist, "Dutch" Stoenner, to try to date the iron meteorites' solidification ages. If it could be shown by any independent method that the irons are the same age as the stones, everything would be fine; if it could be shown that they are a different age, the whole lead-lead dating system would fall apart, and with it all our ideas of solar system formation.

That sounded like fun to Joe Zähringer, so he worked out a method of dating the irons by a variation on the standard method of K/Ar dating, involving activation analysis because the amounts in irons were so low. It would work, he thought, if the argon didn't leak out over time, and if they could measure the extremely small amounts present in the irons.

The first "if" would be tested by the results: low ages could mean diffusion loss and would be meaningless, but ages of four and a half billion would be strong confirmation of the model. The second "if" was tougher: there is very little potassium and even less argon in the irons, and there is a lot of ^{40}Ar in the earth's atmosphere and potassium in the earth's crustal materials, leading to possible contamination and making it impossible to use the mass spec. But, Zähringer hoped, activation analysis would solve both of the second "ifs": upon irradiation, potassium forms ^{42}K, an isotope with a 12.5-hour half-life, and argon forms ^{41}Ar, with a 1.5-hour half-life, and these could be measured even at the low levels expected. Before irradiation the meteorite's surface—and any contamination—would be etched away in an acid bath.

Any contamination occurring after irradiation wouldn't interfere, since it wouldn't be radioactive.

So they worked out the details and carried out the experiment on several irons, and the first "if" vanished—the ages were definitely not low, so argon leakage was not a problem—only to be replaced by an "Oh no!" The data indicated ages of ten billion years for the iron meteorites.

∽

Nobody believed it. If it was just one meteorite, you might just possibly imagine it had originated in another solar system, but they measured several and got roughly the same age for all of them, about ten billion years, and the distances between stars are just much too great for frequent visitors. *On the other hand,* the solar system just couldn't be old enough to accommodate the iron ages. For one thing, how could the irons be some five billion years older than the earth and the stone meteorites? For another, ten billion years was older than the galaxy! And for another, such an old age would screw up everything we knew about the formation of the elements, since, for example, extrapolating the ratio of ^{235}U to ^{238}U back that far meant that more than 99% of the uranium would have been formed as ^{235}U, and that (according to the nuclear astrophysicists) just could not be.

On the other hand...

On the other hand, I thought, you have to go with the data, don't you? The Russian lead data were probably just contamination, and the helium in Washington County was probably not radiogenic at all. But still, the only direct measurements of iron meteorite ages gave ten billion years, not four and a half. And though we thought the galaxy was only 9.7 billion years old, how well did we know that number? Not very well, I thought. And the nuclear astrophysicists were relying on theory, not experiment, for their isotopic argument, and although I wasn't nearly smart enough to argue with them, how about remembering the aether?[9]

What if Stoenner and Zähringer were right?

Visions of sugarplums danced in my head, but I closed my eyes tight and shook them away (nearly driving off the road, evoking screams from my wife and squeals of delight from the kids). Most likely, as I came to my senses, was that the S-Z experiment was wrong, that they had missed something. The first thing to do was get back to Brookhaven and find their lab notebooks.

9. At the end of the nineteenth century, Lord Kelvin, acknowledged *prima* of the world's theoretical physicists, had stated publicly that "If there is one thing we are sure of concerning the universe, it is the existence of the aether." And just a couple of years later, along came Einstein . . .

TEN

K/Ar and the Irons

The world is a complicated place, Hobbes.

—Calvin

ONE DAY AT ITHACA I HADSCREWED MY COURAGE to the sticking point, hopped on my Honda scooter, scooted over to the Ithaca airport, and joined the East Hill Flying Club, an organization that owned a Piper Cub and a Tri-Pacer, and I learned how to fly. I had taken a few lessons at the age of fourteen, but quit when we began to do stalls and my stomach had dropped faster than the plane. Now I found that although I was still scared, I could handle it, and I progressed quickly. Probably the single most terrifying, exhilarating moment in my life was my first solo.

I hadn't yet earned my private pilot's license, but I was able to fly by myself and was allowed, even encouraged, to take short cross-country trips. For this—and for me—Ithaca was ideally suited. The Tri-Pacer had a four-hour range at 120 knots cruising speed, and Ithaca was well within flying range of Washington, New England, New York—and Brookhaven. I took off and was soon approaching Long Island Sound, and having second thoughts.

Whenever I flew out of sight of the Ithaca airport I not only continually looked around the skies to be sure there were no other planes anywhere near me, I also kept my eyes on the ground, picking out level places where I could put the plane down if the motor in front of me ever quit. Now, approaching the Sound, it looked vast and never-ending, with Long Island nothing but a dim, dark line on the horizon. If the engine quit over that water, if I went down...

I turned around, was ashamed of myself, turned back again, turned around again, took a deep breath and headed out over that endless expanse of water.

Ten minutes later I was approaching Long Island. I skimmed over Port Jefferson, found the little airport that served the lab, and set her down smoothly. A cab took me to Brookhaven, I said hello to everyone, found Joe Zähringer's notebooks, and was amazed.

He hadn't taken into account the cosmic rays or the possibility of primordial argon. (The basic assumption in potassium-argon dating is that when the meteorite solidified—the time you're trying to date—there was no argon in it, so that all the argon you measure is the result of potassium decay since that time. In essence, the clock starts ticking at that time. But if the assumption is wrong, if the rock started off with argon in it, then the clock had never been reset to zero; or if the argon is being produced by cosmic rays as well as potassium, then the clock is ticking at a different rate. Either way, the date you measure is too old.)

Luckily, it's an easy correction to make. In fact, I was astonished that he hadn't done it. All you have to do is keep measuring the argon in a Geiger counter for a few months: as the short-lived ^{41}Ar (which gives you the ^{40}Ar abundance) dies away, the radioactivity settles into the 35-day decay of another argon isotope, ^{37}Ar (a measure of the ^{36}Ar abundance), and ^{36}Ar in the meteorite either is produced by cosmic rays or it is primordial or atmospheric. By taking each possibility separately into consideration, you can set limits to the amount of primordial or cosmogenic ^{40}Ar and thus correct the age.

I didn't really believe the ten-billion-year ages, but if they did turn out to be correct—if by doing the experiment properly I could convince the scientific community that everything we knew about the origin of the solar system was wrong—

I didn't really think that was going to happen. But even if it turned out that the high ages were due to cosmic ray or primordial argon, it would be interesting. It was a good experiment to do, so I flew home a happy little bugger, and I wrote a new proposal to the AEC, and I got funds to go somewhere and do the right experiment the following summer.

WARNING: *This chapter is a tale of things gone wrong and ideas that didn't pan out. So if all you're interested in is learning about the universe, skip it. On the other hand, if you want to know how science works, in fits and starts, this is about as good a description of one of the fits as you're likely to find.*

I decided to go back to where it all began, to Brookhaven. The lab made provisions for summer visitors, housing them in an on-campus compound of small wooden shacks. With a playground right in the compound for the kids and Westhampton Beach close by for weekend fun, and with all the facilities of the national lab right there, it made a wonderful place to spend a few months. So I wrote to the man who had the lab next door to Ollie Schaeffer, Ray Davis, who happened to be an expert in the measurement of the radioactive argon isotopes.[1]

Ray was happy to see the Stoenner/Zähringer work followed up, and he invited me to spend the summer in his lab. We moved our family into the rather primitive but clean and pleasant cottage on site, and I quickly put together a vacuum line in Ray's lab and got to work. I followed the Stoenner/Zähringer method (with my chemistry background I was able to do both parts of it): the meteorites were irradiated in the reactor, inducing the $^{40}Ar(n, \gamma)^{41}Ar$ reaction, which enabled me to measure the ^{40}Ar content by counting the ^{41}Ar radioactivity; the $^{41}K(n, \gamma)^{42}K$ reaction, which gave me

1. Davis' work on neutrinos is discussed in chapter 17.

potassium by counting the radioactive ^{42}K; and finally the $^{36}Ar(n, \gamma)^{37}Ar$, which provided the long-lived ^{37}Ar radioactivity.[2] I melted the meteorite in the vacuum line, purified the argon, and pumped it into a Geiger counter. While it was counting, I washed down the inside of the crucible and all the glass around it with hydrofluoric acid to dissolve the potassium, purified it chemically, and put it into another counter. Because of the mixture of half-lives involved, from ^{41}Ar's 1.8 hours to ^{42}K's 12.5 hours, and because we had no automatic counting equipment in those halcyon days, I had to count and record continuously for a couple of days.

The lack of sleep almost killed me. Literally. The lab's air-conditioning wasn't equal to the summer's humidity, and periodically I had to turn off the Geiger counter, which was running at 2,000 volts, take out the connections, and wipe them down with acetone to dry them off. One night—or day, I don't remember, they all melted into each other in the lab—well into the second twenty-four hours without sleep, I forgot to turn off the counter before reaching behind it and taking hold of the connector—

The next thing I knew I was sitting on the floor on the opposite side of the room with my legs splayed out in front of me, shaking uncontrollably. I'm just lucky the shock threw me loose.

The other incident was more subtle. After one particular experiment, I woke up from a long sleep and found, as I was eating breakfast, that my finger was itching. By the next day there was a definite white spot and the itching was worse, so I went into the infirmary where the doctor, after scratching his head, asked what sort of experiments I was doing. He was evidently accustomed to sloppy chemists. When I mentioned the hydrofluoric acid he smiled happily. That was it, he said, you must have spilled a drop.

What do we do? I asked.

He shrugged.

2. The $X(n, \gamma)Y$ terminology means that X is hit by a neutron (n) and forms Y with the release of a gamma ray (γ).

The experiment involved radioactivity, I suggested, and he panicked.

"We'll get you on a helicopter to Rochester!"

"What?

"They have a radiomedicine facility there!"

"We don't have one here?"

"No!" He was already dialing.

"Brookhaven doesn't have—"

"No! We can't handle it!"

I didn't want to go to Rochester and leave my family here. "Wait a second," I lied, "I just realized the HF wasn't radioactive, we were finished with that stuff by then."

He hung up, relieved. He didn't even question me about my volte-face, he was just happy he didn't have to deal with it.

"So what do we do about the HF?" I asked.

"Not much to be done. We'll swab it down, but basically we just wait."

"For what?"

"Depends on how much HF you got on you. If there was enough, it'll eat through your flesh to the bone, and then we'll amputate. If there was less, it'll stop before it reaches the bone."

For the next few days the white spot deepened into a depression, then a hole, then a deeper hole—and then it stopped, and my finger stayed on.

Aside from that, the summer was fun ("Yes, yes, but aside from that, Mrs. Lincoln, how did you enjoy the play?"), and the results were even more fun: I measured a variety of iron meteorites, and neither the cosmogenic nor the primordial argon altered the results in any appreciable way. One iron meteorite gave a young age (1.7 billion years), but all the others really did seem to be about ten billion years old. But because accepting such an age as real would be a tremendous step to take, I had to nail it down.

The next step was to show that the argon followed the potassium wherever it happened to be in the meteorite. For this I needed a nice clean iron crystal, and I found the perfect specimen in the Santa Rosa meteorite. Most irons consist of multicrystals with cracks and imperfections that might allow the argon to migrate away from where it was formed, in essence destroying the evidence. But Santa Rosa was an ataxite: a single iron crystal with no pathways for migration. If the argon really was a decay product of potassium, then it would be trapped in the crystal right where it had been formed ten (!) billion years ago, and I would find it there today.

I got a large sample from the New York Museum of Natural History and was ready to astound the world. All I needed was a little bit of luck. And I very nearly had it.

My first sample went well. There was a lot of argon, so after I finished the potassium chemistry and was ready to put it into the Geiger counter I could predict a lot of potassium, and it was there; the age for Santa Rosa was typical of the other iron meteorites, 10.3 billion years. The next sample had only half the amount of argon, so I was expecting—hoping—for half the amount of potassium, and that's what I got: the age was 10.5 billion years.

This was it! A good variation in the radioactive potassium, accompanied by the same variation in the daughter argon. This was the kind of correlation that was needed to prove the case. I went home on my scooter in wild abandon, imagining great glories, and the next morning hurrying on my way back to the lab I hit a dog and the scooter flipped and I flew through the air—

I woke up lying on my back in the street, surrounded by a circle of peering children. As I opened my eyes one of them snorted in disgust, "Ah, he's not dead!"

Neither, to my own disgust, was the dog. He was a small, snarling beast that belonged to a physicist friend of mine, Bruce Hapke, who always came barking and yapping at me as I drove down the street (the dog, not Bruce), and this time as I tried to avoid him he had darted in front of my bike and that's when I hit him. But a Honda scooter at maybe twenty-five miles an hour isn't a Harley-Davidson at seventy, and the little beast didn't even have a broken bone.

On the other hand, neither did I. The Honda was dented but still rideable, and so I got back on and went to work—and did one experiment too many. For as it turned out, the accident wasn't the worst thing that happened to me that day. The next sample of Santa Rosa had a small amount of argon—but no potassium at all! When I put it into the counter there just wasn't anything there. Nothing. Nada.

Well, I thought, I must have goofed on the chemistry. The procedure involved measuring the argon first, because of its short half-life; it took about five hours. Then I had to wash out the glass and the crucible with acid (HF, and now I was wearing plastic gloves) to dissolve the potassium, and then go through a series of complex precipitations and dissolutions to purify it. I would precipitate it and throw away the solution, keeping the precipitate. Then I'd dissolve it and precipitate any iron left. I would throw away that precipitate and keep the solution, then reprecipitate the potassium and discard the solution, etc., etc., etc. So the chemistry was arduous, taking another ten hours or so and involving the throwing away of alternately precipitate and solution. If I just once threw away the wrong stuff—kept the iron precipitate, for instance, and threw away the potassium solution, I'd lose it all.

I didn't really think I had done that, but by the time the potassium was ready to be counted I had been working more than twenty-four hours and I was tired, and maybe—hopefully—I had goofed. I took a day off, slept late, and then went in for another run. This time the argon was midway between the other values, but the potassium was much lower, giving a high age, and the next day the argon was lower but the potassium was higher. . . .

The correlation was gone. I kept on at it, but the ages were all over the lot.

They were meaningless.

∞

Well, what does "meaningless" mean? They had to mean something, but I didn't know what. I published them as "anomalous," and didn't know what to do next. When I told Ray about this he was as upbeat as possible. "Come back here next summer," he said, "and we'll figure out what to do next. In the meantime..."

I knew what to do in the meantime. From September till May I taught my courses and went over and over the data, searching for clues, finding nothing. Then, one morning in early May, I picked up a letter from Brookhaven:

> Dear Dr. Fisher,
>
> I am sorry to tell you that we will not have room for you at the laboratory this summer. Best wishes for your future work.
>
> Sincerely,
>
> R. Dodson.

R. Dodson was Dick Dodson, the chairman of the chemistry department there. I couldn't believe what I was reading, I couldn't believe that Ray would renege on his invitation...

He hadn't. He called a couple of weeks later to say he was all set up for me, and when was I coming? I told him about Dodson's letter, and there was a long silence, and then he said he'd call me back. When he did he was quietly furious. The letter was no mistake. The senior chemistry faculty at Brookhaven evidently had decided, he said, that his work on neutrinos was a mistake, a waste of time and of departmental facilities. They couldn't fire him since he had tenure, but they could put pressure on him to quit. The first thing they decided to do was to take away his privilege of bringing in summer scholars. They didn't announce this, of course; they didn't want to get into a tenure battle. They just decided to decide, case by case, that whomever he invited was to be disinvited with a noncommittal letter.

What the hell. Although I thought they were shits for the cowardly way they were going about it, I didn't really disagree with their evaluation. I too thought his neutrino work was a waste of time, so I tossed the letter and got on with my life.[3] By this time I had attracted several graduate students, and one of them, Lou Rancitelli, had worked out a better experimental procedure: instead of melting the meteorites in order to get the argon out of them, we'd now dissolve them in acid. This meant a tougher job cleaning up the argon, because of the chemical reactions taking place in the acid, but Lou worked out a good procedure. And now, instead of dissolving the potassium off the walls of the glass and the crucible, we had it already in solution and ready to go.

But we didn't know what meteorite to study. Nothing looked like a better bet than Santa Rosa, and that hadn't worked out. There was no point in just measuring more "ages" when we knew that the ages weren't real. And then Gerry Wasserburg came into my life.

G. J. Wasserburg is one of the top geo/cosmochemists in the world, but a tough man to deal with on a personal level; well, a son of a bitch, really. He was a full professor at Caltech (I was still an untenured assistant professor at Cornell) who would a few years later establish the premier laboratory for the isotopic study of lunar samples (he named it the Lunatic Asylum) and win just about every award available to a geoscientist. His name now leaped off the pages of a newly arrived copy of the journal *Science*. He and his coworkers had managed to separate some silicate inclusions from the Weekeroo Station iron meteorite, and had measured its age in the manner of normal stone meteorites and got a value of 4.6 billion years. They concluded that the K/Ar ages Joe Zähringer and I had measured on the irons were invalid: the irons and the stones had solidified at the same time from the same primordial material.

But during my Brookhaven summer I had found a young age in one iron. So Wasserburg's conclusion that their work invalidated the ten-billion-year iron ages would be valid only if Weekeroo Station showed a ten-billion-year K/Ar age, as did most—but not all—of

3. And it's been a good life, but not a Nobel one. And no one at Brookhaven besides Ray Davis has ever won a Nobel Prize. His name will live in science history, while I'm probably the only person who remembers the name "R. Dodson."

the irons. I wrote to him, explaining this and asking for a sample so I could do the K/Ar age determination. I got a letter back from his colleague, Don Burnett, saying that Dr. Wasserburg was on sabbatical in Switzerland, but that he (Don) had contacted him about my letter and they agreed that I should do the experiment. A sample of Weekeroo duly arrived and Lou Rancitelli and I got to work.

One of the necessary ingredients, as I have said, is determining the cosmic ray and/or primordial contributions to the potassium and argon, which meant measuring the other (nonradiogenic) argon isotopes. As it turned out, they made a very small contribution to the age, which was the "normal" ten-billion-year result. Since the silicate inclusions were only 4.6 billion years old, there was now no possible doubt that the K/Ar ages we had been getting for the irons were just plain wrong.

Rancitelli and I wrote up the results and sent them off to Wasserburg before sending them to *Science* for publication—and I got back a vituperative letter from Gerry, accusing me of everything from bad manners to a lack of scholarly conduct. It turned out that he was now working on cosmic ray effects in meteorites and felt that I had scooped him on his own sample by including the cosmic ray measurements on Weekeroo Station, which he felt he had a proprietary right to. I wrote back that he must have known that we had to make those measurements in order to establish the K/Ar age, and that we hadn't used the data for anything else related to cosmic ray effects: we didn't try, for example, to calculate a cosmic ray age, or compare the data to other meteorites.[4]

Gerry never answered my letter. Severe repercussions would follow.

4. In an autobiographical article, "Isotopic Adventures: Geological, Planetological, and Cosmic" (*Annual Review of Earth and Planetary Sciences,* 31 (2003): 1–74), he writes about a different set of events earlier in his career, when he was evidently on the other side of the argument: "Proprietary rights to the chart of the nuclides was not understandable to me, nor were field rights to geological regions . . . Over the years at Caltech, I have had three Division Chairmen come to my office and inform me that I should not work on some element or isotope because they 'belonged' to a colleague. Similar things happened later in another field, when some stars or stellar processes were considered to be personal property. Well, *ad astra per aspera*." To which I can only reply, "*Et tu, Brute?*"

ELEVEN

Interlude: The Spreading Oceans

Nice work if you can get it . . .

—Ira Gershwin

WELL, THE STORY OF THE K/AR AGES OF IRON meteorites ended not with a bang but a whimper. We were at a loss to explain them, so for the moment we concentrated on tuning up the experiment, trying to find some error in our technique.

We couldn't. Yet clearly something was wrong. By this time Rancitelli had measured a large number of meteorites, so we thought of plotting the measured ages against the argon contents: if there was any truth at all to the ages, they should be proportional to the argon.

They weren't. The data showed nothing but scatter. Just for fun we plotted the ages against the potassium content—and there was a definite anticorrelation! No question about it, there it was: the more the potassium, the lower the age. Not only that, but the extrapolated end point at the high potassium end gave an age of just about 4.6 billion years! The whole story—the high potassium-argon ages, the ten-billion-year story—was due to leakage of potassium from the meteorites.

It had never occurred to us. We had expected that if anything, during weathering on earth and during our cutting and cleaning of the meteorites, they might lose argon, since argon is a gas and loosely bound; if they had, the true ages would be even higher than our measured values. Now we realized—well, hypothesized—that since the argon is formed in a radioactive decay it comes with a burst of kinetic energy and might well lodge itself in the iron matrix, where it would stick while the potassium slips out.

Monday morning quarterbacking. Hindsight.

So there went five years of my life. Trying to measure the potassium-argon ages of iron meteorites turned out to be a useless endeavor, a waste of time.

Damn you, Stoenner and Zähringer![1]

All in all, it wasn't a great year. I was finishing my fifth year at Cornell, and was due to be either promoted and given tenure or kicked out. I wasn't too worried, since my primary appointment was in a small department, Engineering Physics, and my research record was just about the best of the department.[2] If I had been in the physics department, I wouldn't have expected tenure, because the physics department at Cornell was one of the best in the nation and I wasn't up to that standard, but for where I was I was doing well. So I wasn't surprised when one afternoon as I was leaving campus to fly my little Piper up to Hartford with

1. The only thing I got out of it was one last publication, and one last aggravation. We sent the manuscript off to the *Journal of Geophysical Research*, where it languished for several months. When I called to ask what was holding it up I was told one referee wasn't replying. Finally he did, with not a single comment to make. So what had taken him so long? It was Joe Zähringer, and we soon realized why he had held on to it. He was working along the same lines, and while he held our ms. he sent his own in to another journal, where it appeared first. Well, what the hell. I guess he felt he owned the K/Ar story; he had, after all, started it.

2. Aside from the K/Ar stuff, I had pioneered the use of instrumental activation analysis, applying it to stuff from cows to rocks and meteorites, and had done some analytical work on meteorites, and was publishing a good deal. Solid work, though nothing really important.

my wife and kids, dropping them off with my mother-in-law there before continuing on to New Hampshire for a conference, I was stopped by the head of the aeronautical engineering department. He smiled and shook my hand and said he had just come from the meeting of my tenure review committee, and they had unanimously granted me promotion and tenure.

So that was nice, I thought, as I flew away for a week, although I wasn't hysterically happy because I wasn't sure I wanted to spend the rest of my life in Ithaca. I remembered my first day there, when my wife and I had stood on the hill at the top of the road leading down into Main Street, and we looked at that dreary town—you could see the whole town from this spot—and we looked at each other and tried to smile. We just couldn't. In Florida we had had the beaches, at Brookhaven we had had not only the beaches but just a short drive into the City... here there was nothing but a little town with one movie theater, no live theater or decent restaurants, no public television, and hundreds of miles from anything that looked like civilization. "Centrally isolated," Hans Bethe famously called it.

We took our children (soon there was a third) to the Ithaca zoo. It had a dog, a chicken, and something that looked like a mole or ferret. We went to the "beach" at the foot of Lake Cayuga, where the dirt beach was covered with scum and fuel oil debris from the rich people's motorboats.[3] We went to a movie on Saturday night and sat with a house full of noisy students, then drove home through dark quiet streets because there was nowhere else to go.

Ithaca was not a fun place unless you had friends. I was the youngest man in our department by about forty years, so that

3. The public beach is at the south end of this finger lake. The northern shores are lined with the homes of the rich, each of whom seemed to have a motorboat, and God in his wisdom had provided the region with northerly winds which skimmed the surface and brought the accumulating oil scum southward and deposited it on the beach. Our first day at the beach was misery; later, when we became friends with Tommy Gold, who had a motorboat, we loved the water skiing and forgot about the oil scum and the poor folk.

avenue was out. Our first year we made one friend, a postdoc in physics, but he left at the end of the year. Our second year we buddied up with a postdoc in biology, and at the end of the year he left. Our third year it was another postdoc in physics, and again by summer he was gone. We came back to Ithaca after summer vacation and I thought, we've lived here three years and we have no friends and we do not like this place; I have to find another job.[4]

But that year it all changed. Tommy Gold, the astrophysicist who originated steady state theory and explained pulsars as neutron stars (and had a motorboat), was head of the astronomy department and the Center for Radiophysics and Space Research. He invited us for dinner with the Phil Morrisons, and first off we were taken down to the cellar, where he had a rope hanging from the ceiling; he climbed it legs-free and then invited me to take my turn. I flunked that test, did only slightly better on the balancing board and the finger coordination game, and totally disgraced myself when we turned to rapid sit-ups. Morrison, a polio survivor, was exempt from the games but came into his own upstairs over drinks, where I was blown away by the heady intellectual conversation. I didn't distinguish myself there either, I just sat and listened as they discussed all the marvelous things we could do now that we had unlimited energy,[5] and I thought we'd never be invited again. But Tommy wasn't too much older than we were, and it turned out that my wife, Leila, had charmed him, and we became frequent visitors at his house.

4. The entering freshmen felt the same way: their morale was low. Almost none of them had picked Cornell as their first choice; they had all been rejected at Harvard, Yale, and Princeton. One of the first warnings I had been given as a faculty member was to try to be gentle with them; too many were jumping off the bridge at "Suicide Gorge," although within a couple of years they would generally begin to recognize what a great university it was.

5. This was 1963, and the vistas for nuclear energy were unlimited. And then Greenpeace and the folk singers confused nuclear energy with nuclear bombs, and confused the public with their rhetoric and songs, and the vision disappeared in the muck of carbon dioxide that has given us global warming.

At one of his parties Peter Hilton, one of the world's foremost topologists, tapped me on the shoulder and introduced himself. "I understand you and I have something in common," he said. I tried to think of something suitably brilliant that I had done, but he went on: "We have the only beautiful wives in Ithaca," he said. It was true. Leila looked like a cross between Sophia Loren and Jackie Kennedy, and Maggie Hilton was a lovely actress. The four of us quickly became close friends, and with them our circle enlarged. For the next few years there were frequent parties with wonderfully intellectual people and lots of dancing, where I would drink until midnight, when the cake and coffee came out, and then I'd drink coffee for the next two or three hours, and finally drive home carefully and sleep soundly.

When Maggie and a few others at Cornell and in town revived the amateur acting group called the Barnes Players, I joined, and it became a pivotal part of my life. So the last few years had been good, despite the noble gases, and yet Ithaca loomed dark and gloomy as a place to spend your whole life.

I didn't have to worry. I nearly didn't have a whole life left. Returning that year in my Piper Tri-Pacer from the conference in New Hampshire, I began to run into lowering clouds. I didn't have an instrument rating—had never flown on instruments—and so I flew lower and lower to keep under them. But they just kept on lowering, and the hills in front of me started rising, until finally I was below two thousand feet and my sectional map told me there were peaks higher than two thousand in front of me.

I should have turned around and gone back. Certainly I should have landed somewhere and waited for the clouds to lift. But my wife hated flying in that little thing and I had convinced her to do so only because it was such a convenient way to get around New England, and if I had to call and tell her I wouldn't be back today because of the weather and all she had to do was look outside and see the weather looked fine except for a few clouds—

I kept on, and finally I was down to one thousand feet and the hills really were reaching into the clouds and I just pulled back on the wheel and the Piper's nose lifted and seconds later we were flying in a cold white floculence that was bumpy and blinding, and in another few seconds I didn't know up from down from sideways and I felt the panic rising in my gorge, choking me—

I took my hands off the wheel and pulled back the throttle and trusted the plane to fly properly by itself. The nose dropped and it fell away and it was all I could do to keep my hands off it until a few thousand years later we dropped slowly out of the clouds and I could see again. And the Lord was in his heaven and everything was all right because it looked lighter up ahead so I kept going and the clouds began to lift and the hills fell away and I flew back to Hartford and my wife never knew about that little moment of uncertainty (as I now began to think of it).

But when I finally returned to Ithaca I found a letter waiting for me from the head of our department, informing me that I had officially been denied tenure and should start looking for another job immediately.

What the hell? As soon as I learned I was being kicked out I wanted to stay. But what had happened? Jim Howe, chairman of our department, wrote to me, "There seems to be a feeling among the faculty that you don't work hard enough. You get a lot done, but you don't seem to spend a lot of time doing it."

Huh? I think I know what they were thinking: I was publishing short stories in magazines with titles like *Dude* and *Gent*, I was spending many afternoons either flying or skiing, and I was spending evenings acting with the Barnes Players. I guess I didn't give the impression of seriousness that an Ivy League school wanted. But there had to be more to it than that.

I didn't understand it then, and I'm still not sure of the whole story today. But one way and another it turned out to be a classic story of university politics, with several subplots.

Subplot No. 1. Main character: Tommy Gold. Anyone who ever met Tommy knows that he was[6] brilliant, athletic, charming, and a pain in the ass (not necessarily in that order). But perhaps his dominant trait, for this particular story, is "threatening." To the engineering faculty his force of personality was so strong, his intellectual dominance so overwhelming, and his disregard for lesser mortals so obvious, that he terrified them. His Center for Radiophysics and Space Research was an interdepartmental research center in the college of Arts and Sciences, but it was innately strong in engineering and it was quite successful. It had its own building (known to the grad students as Uncle Tom's Cabin) and a long string of research contracts and postdocs.

Several upper-level people in the engineering school who were predisposed toward me let me know that the deanery there was afraid that Tommy was trying to move into their school. It would have been good for the engineers if he had, but you know the old saying: first-rate people hire first-rate people; second-rate people hire third-rate people. Tommy was no threat to the physics people, but the engineering school wasn't up to that standard and they were afraid Tommy would take them over. I was seen as one of his "people," the wedge in the door.

So they shut the door.

Subplot No. 2. Main characters: the engineering administrators. One of our best friends at Cornell was the astrophysicist Ed Salpeter, who volunteered to look into the situation for me. He came back with a somber face: in my file was a letter from Jim Arnold, who had been one of the leaders in meteoritics a few years back. Ed said that he had been told the letter said, quote: "I cannot recommend David Fisher for tenure . . ."

I really was astonished. I couldn't believe Arnold would say that, and I asked Ed if he could find out more. A few days later he said he had read the letter, which actually said, "I cannot recommend

6. Tommy died in 2004.

David Fisher for tenure because I am no longer in the field and have not been keeping up with the literature. I suggest you contact Ed Anders at Chicago." And there was a very complimentary letter in the file from Anders.

I was angry, but Ed said to let it go. "If they really don't want you, the reasons don't matter, you won't be happy here." If anything like that happened today, the bastards would be sued. But I took Ed's advice. I let it go.

Subplot No. 3. Main character: G. J. Wasserburg. I don't know how it happened, but somehow they had got hold of Wasserburg's letter to me complaining of my "lack of scholarly conduct." This was damning, coming from such a notable scientist. I thought at the time he must have sent a copy to my dean, which would have been incredibly nasty. But years later Cesare Emiliani, who hired me at Miami, told me that the deciding factor in their offering me a position was a laudable recommendation from Wasserburg. So go know.

Subplot No. 4. Main characters: Tommy Gold and Leila Fisher. Tommy had been having an affair with the wife of one of my colleagues in the engineering school. He had grown tired of the affair, and since she and her husband were off for a sabbatical year in Europe, he was hoping the whole thing would peter out and be over by the time they returned. In the meantime, he met and was charmed by my wife, who flirted with him quite innocently; she was from Hartford, and it never entered her New England–bred head that people would actually sleep with each other outside of marriage.

So then the sabbatical year was over and the woman returned to find Tommy not only cold to her but obviously pursuing another woman; in fact, she assumed they were sleeping together. She told me so at one of our parties, as we were dancing, and suggested that the two of us should get together. I was also pretty innocent at the time, and was shocked—not at what she said about Tommy and Leila, for I knew Leila would never sleep around, but at the thought that I should "get together" with her. I didn't know how to handle it, and I handled it badly: I laughed. My intention was that she should think that I thought she was joking and so she could laugh too and she wouldn't be embarrassed at being refused, but she took it the other way: she thought I was laughing at her, and I have never seen a woman so suddenly and so violently angry. The effort at suppressing it brought the blood rushing to her face as she broke away from me.

I didn't know what to do, so I did nothing. And I heard no more about it for three years, until Tommy, now trying to find out why I had been refused tenure, called her husband, my "colleague," who told him one half-truth and one violent lie.

The half-truth: The day I wanted to do several runs on the reactor to measure aluminum, I had asked the student who was scheduled to work that day if he minded if I took his place. It was a Saturday, and he was happy for the excuse to leave. Now Tommy was told that I had "bumped" him and the student had complained that I had pulled rank.

The lie: Another weekend I had gone in to work but the multichannel analyzer wasn't working properly. I didn't know how to fix it so I left a note detailing the problem, and went home. Now Tommy was told that I had broken the machine and hadn't told anyone, "hoping no one would know it was [him]."

In Albee's *Who's Afraid of Virginia Woolf*, George complains that in order to succeed in academia you have to "plow a few pertinent wives." I hadn't plowed the impertinent wife, and George was right. So, all in all, the conditions at Cornell weren't favorable, and I began to look elsewhere for another job.

And again ran into university politics.

ꟹ

Ed Fireman called Fred Singer and told him I was "available." Fred had just been hired at the University of Miami to create a new School of Environmental and Planetary Sciences (SEPS). In 1965, this was way out on the bulging forward edge of the envelope, and would have put UM in the lead for several new areas of coming research. We had just put up our first satellite a few years earlier, so space research was set to take off, and environmental science was just hearing the first teeny little bits of hints about the possibility of global warming.[7] So the administration at UM had a good idea on their hands, but little money to back it with and not the slightest hint of what to do with it.

Over the Thanksgiving break, Fred invited me to visit, putting me up at the Fontainebleau Hotel on Miami Beach and taking me to the Playboy Club for drinks and dinner. It was twenty degrees below zero when I left Ithaca and seventy-five above when I arrived in Miami, and I began to bless the witches who had cursed my tenure at Cornell, until we finally visited the UM campus, which was pretty, but scholastically bleak. I could have a joint appointment in chemistry or physics, Fred said, but when I talked to the faculty—well, let's just say they weren't up to the Cornell standard. And when we walked over to the building where SEPS would be housed, it was just an empty shell. There were no lab facilities and no plans (i.e., no money) to build any in the immediate future.

I felt like Cinderella at one minute past midnight. And then we drove out to the Marine Lab on Virginia Key, and someone turned back the minute hand. I met Fritz Koczy, Cesare Emiliani, Joe Prospero, and Göte Östlund, and felt right at home right away.

The Marine Lab was one of the three or four top oceanographic institutions in the country. It was to be incorporated in SEPS as a distinct but subordinate entity, rather equivalent to a normal university department, and Fred—seeing the immediate

7. The environmental movement had just begun with ill-founded hysteria over nuclear energy and DDT, and had begun to swing over into climate worries. The earth's temperature had seemed to be cooling over the previous few decades, and worries that pollution might speed our descent into a new ice age were bruited about. At a conference at Brown University, called to discuss global cooling, one paper suggested that the carbon dioxide that accompanied most pollutants would instead warm the planet...

rapport I felt with these guys—said I could have a joint appointment with them instead of with physics or chemistry on the main campus. I said I'd want to have my primary office and lab out here on the Key rather than in the empty SEPS building, and Fred thought that would be fine. I went back to Cornell a happy man, to await the formal offer.

And I waited, and I waited...

Finally the envelope arrived. I tore it open, and read the two sentences. "I am sorry...." and "Best of luck in your future endeavors..."

Shit.

I called Singer, who was apologetic but said it just hadn't worked out. The people at the Marine Lab objected to giving me an appointment there; they didn't want to work with me, and there was nothing he could do about it. That really hurt. I thought I was a pretty good scientist and that we had gotten on well together, but, well, the hell with them.

In my activation analysis work I had collaborated on a couple of projects with a scientist at General Atomics named Roman Schmitt. He now accepted an appointment at Oregon State University, and wanted to hire me to work with him on analyses of the soon-expected lunar samples. Leila wasn't happy about moving to the other side of the continent, and although the lunar samples were intriguing, I wasn't happy about further activation analysis work. To be honest, I didn't know what I wanted to do anymore. But the months dragged by and nothing else came into view, so I tentatively accepted Roman's offer. I put off a final decision, blaming it on my wife—"She's concerned about moving so far from home"—and told him I'd sign the contract right after the AGU meeting.

The American Geophysical Union annual meeting in Washington, D.C., took place in April, and I was hoping there might be something happening there. But there wasn't, and on the final Friday I checked out of the Statler and started to head home, psyching myself up for the move to Oregon. As I walked out the door of the hotel and headed to the cab stand I heard a shout, and turned to see Cesare Emiliani waving at me. He had been so friendly in Miami, but then—I didn't want to talk to him.

But he came rushing over, all smiles. "A man can smile, and smile, and be a villain," I thought, hardly listening as he said, "When are you going to come?"

I looked at him. "What?"

"No," he said. "When?" He laughed.

"I don't know what you're talking about," I said.

"We got rid of Singer," he said. "Now we want to hire you. The oceans are spreading!"

What had happened was that the UM administrators had been their usual dense selves. Without a clue. They had hired Singer to be dean of this new school, SEPS, but the only part of it that had anything worth having—faculty, grad students, lab equipment, reputation—was the Marine Lab. Imagine how the people at the Marine Lab felt when they were told that this big shot physicist was now their boss. It was the classic tale of the tail being hired to wag the dog.

They wouldn't have any of it. They got together and blocked everything Singer tried to do—including hiring any new faculty, which included me—until finally the UM administrative idiots got the message, and instead of rearranging their idea and incorporating SEPS into the Marine Lab rather than the other way around, they simply gave up and canceled SEPS. By the time of the AGU meeting, Singer had left UM, and now the Marine Lab wanted to hire me. Because, as Cesare said, "The oceans are spreading!"

Literally. Which takes us back to argon.

TWELVE

Dating the Spreading Seafloor

A delightful chat of this and that
. . . and cocktails for two.

—A. Johnston & S. Coslow

IN 1912 A GERMAN METEOROLOGIST, ALFRED Wegener, took the drastic step of moving into another science altogether by publishing the shocking geologic theory that our continents have been sailing across the surface of the earth like leaves on a lake blown by—what? The geologists laughed at the suggestion of an impossible wind and scorned the man who had insolently crossed the boundaries of the sciences.

But truth be told, it wasn't unheard of in those early years to do just that. Rutherford, a physicist, had won the Nobel Prize in chemistry, and Marie Curie had already won twice, once in physics and once in chemistry. Wegener himself had done his PhD work in astronomy before switching over to meteorology, and at the same time was a renowned arctic explorer. The separation between the sciences are useful and real—a biology student has enough to learn without spending years on tensor analysis or relativity—but at their boundaries they blur. Today nearly everyone pays lip service to what we call interdisciplinary research, but in practice they fight hard against it. I did my PhD course work in the chemistry department of the University of Florida and my research in the physics group at Oak Ridge, then had a postdoc appointment in the chemistry department at Brookhaven before going to physics at Cornell and ending up in geology at Miami, but I had to fight along the way. A chemistry professor at Florida tried to insist that I take his colloid course instead of relativity (which was taught at the same time). I won that fight but lost at Cornell when I tried to have my students take chemistry courses instead of the required engineering and physics courses.

The fact that Wegener wasn't a geologist gave them an easy way out: it's easier to laugh at new ideas than to confront them, and easier still to laugh at new ideas from those whom you can consider amateurs. And at first glance it was laughable, for what "wind" could blow the massive continents around like leaves on water? Wegener himself didn't know; he said that this would be the most important problem in geology in the twentieth century: "the Newton of drift theory has not yet appeared." For the geologists, that ended the discussion with a snicker.

Yet the problems Wegener focused on would not go away. The geologists had already noticed and dismissed the first point that had caught his attention. Two years earlier he had written to his then girlfriend: "Doesn't the east coast of South America fit exactly against the west coast of Africa, as if they had once been joined? This is an idea I'll have to pursue." But that idea was easy to disparage. Mere coincidence. Even Freud had admitted that sometimes a cigar was just a cigar.

There was more, but again the geologists were not impressed. The fossils, for example. The ancient fern *Glossopteris* and the small amphibious reptile *Mesosaurus* had left fossils in both South America and Africa. Neither could have swum or flown across the intervening thousands of miles of empty ocean. The accepted geologic explanation was that there had existed a continental bridge between these two continents which has since sunk beneath the waters. This was, of course, a rather ad hoc explanation with no evidence to back it up, but it seemed to them more realistic than swimming continents.

Wegener thought not. He argued that the light continental rocks would not permanently rest on the heavier oceanic basalts, but would—if ever they sank—rise again. After his first paper in 1912 he published a book setting forth his arguments, but it was published in German in 1915, and the world had other things to worry about at the time. It was finally translated into English, French, Spanish, Russian, and German; in it he gathered all his arguments, which now included ancient glaciations appearing simultaneously in South America, Africa, and India and geologic evidence relating mountain regions on both sides of the Atlantic. The book exploded on the international community like a water bomb: Splat. Annoying, but soon forgotten. Wegener's name was linked with that of Velikovsky[1]: at best they were silly dilettantes, at worst, charlatans, and the concept of floating continents sank beneath the consciousness of science.

And then, along came Adolf Hitler.

World wars are not particularly good for humanity, but you have to admit they have a remarkably salutary effect on science and technology. Real advances in science stop during the actual fighting, as most scientists are enlisted in the war effort, but that effort brings forth tremendous advances in technologies which, as soon as the fighting stops, spur science onward and upward in leaps and bounds. Radar gave rise to radio astronomy, the synthesis of penicillin revolutionized medical science, and antisubmarine warfare led us to magnetometers.

Originally developed to detect submarines, by the end of the war they were sufficiently sensitive to detect variations in the remnant magnetism of the ocean floor. And they were no longer needed to fight against submarines, nor were the ships that carried

1. Velikovsky, in his books, attempted to give pseudoscientific explanations of the Biblical miracles, such as the sun standing still and the manna from heaven.

them. Suddenly the navies of Britain and the United States had ships and equipment they didn't know what to do with, and scientists took that equipment and turned to the oceans.

As oceanographers were fond of saying, we knew more about the moon than we did about the floors of the oceans on our own planet. Most scientists thought of them as basically blah and uninteresting, vast plains that served only to separate the geologically interesting continents where all the earth processes were played out. Imagine their surprise in the 1950s when the *Vema,* a research vessel operated by the Lamont Geological Observatory, mapped the floor of the Atlantic and found an enormous mountain range splitting the ocean in two. Other surveys found similar mountains at the bottom of oceans around the world, and when it was realized that these mid-ocean ridges form a continuous chain winding around the world like the seams of a baseball—the longest and tallest chain of mountains in the world—it gave rise to a new/old idea.

The ocean floor was no longer flat and uninteresting, and in 1960 it led Harry Hess, a geologist at Princeton, to publish what he called his "geopoetry." He was a clever man, and his psychology was as clever as his geology. By calling it "geopoetry" he deflected the criticisms which were bound to arise for such a speculative attempt, and although clearly he was suggesting a mechanism for Wegener's idea of continental drift, he never used that terminology, and another scientist, Robert Dietz, gave it a new name: seafloor spreading. The idea was that molten rock from the mantle continuously erupts along the center of the mid-ocean ridge, creating new ocean floor as it flows away, pushed by more molten rock moving up from the mantle to replace it and/or pulled as the ends of the slab sink back into the mantle at the deep ocean trenches which border the continents. The ocean floor is thus a moving conveyor belt with new material originating at the ridges, getting successively older as it spreads across the ocean.

It may have been damned clever speculation, but most scientists agreed that speculation—or geopoetry—was all it was. It made for interesting talk over a couple of beers, but until some sort of observational evidence came in to support it that was what it would remain. And then almost immediate confirmation was found.

Iron minerals are a normal constituent of volcanic rocks. Upon eruption as molten lava, their iron atoms are swirling around at random, but as the lava cools, the iron atoms line up in the direction of the Earth's magnetic field, acting like tiny compasses. Studies of this "remnant magnetism" in volcanic rocks had shown previously that in rocks of different ages these little compasses exhibit different directions because the Earth's magnetic field spontaneously changes polarity from N-S to S-N and then back again. Now, with the marvelously sensitive magnetometers developed in World War Two for antisubmarine detection, this remnant magnetism could be mapped over the ocean floor. In less than a year after Hess's geopoetic suggestion, Lawrence Morley, a scientist at the Canadian Geological Survey, noticed a series of magnetic "stripes" that had been mapped in the Northeast Pacific. He sent a paper to *Nature* suggesting that these reflected a series of magnetic reversals that had been caught and recorded in lavas as they erupted onto the ocean floor. That is, at some time in the past the magnetic field was "normal" (i.e., as it is today), and the erupting rock shows this in its remnant magnetism. Time passes, the field reverses itself, and now the erupting lava captures this reversed polarity. As more time passes and the seafloor spreads across the Pacific, each magnetic reversal is recorded in the continuously erupting lava, just as a tape recorder might record alternating sounds.

Nature rejected the paper. "Too speculative," they said. Fit only for "cocktail party conversation." Morley revised it and sent it to the *Journal of Geophysical Research*, and received a similar rejection. He shrugged and gave up.

Meanwhile a new graduate student at the University of Cambridge, Fred Vine (who had never heard of Morley's idea), was thinking that if Hess was right there should indeed be such magnetic stripes recorded in the rocks of the seafloor, and after a long struggle with the new science of computer programming and interpretation, he found strong evidence in magnetic surveys from the crests and flanks of the mid-Atlantic and Indian Ocean ridges. Most convincing was the mirror symmetry of both flanks of the ridges, and with this paper—which *Nature* immediately published—Hess's geopoetry became a suitable subject for serious scientific work.[2] And serious scientists were ready to think about the ocean floor.

Serious scientific study of the world's oceans began after World War Two, and blossomed quickly into the most important part of earth research. The leading institutions were the Big Four: Scripps Institution of Oceanography in California, Woods Hole Oceanographic Institution on Cape Cod, Lamont Geological Observatory of Columbia University (now Lamont-Doherty Earth Observatory), and the University of Miami's Marine Laboratory (now the Rosenstiel School of Marine and Atmospheric Science).[3] Although each one is unique, the UM history is illustrative of how these grew.

In 1940, when the newly established University of Miami was known as Sun Tan U for obvious reasons, or as Cardboard U because of the flimsy partitions that were hurriedly put up to separate classrooms for the five hundred students in the school's only building, a thirty-one-year-old British marine scientist who

2. Morley had submitted his work twice and it was rejected twice. The reviewer's remark that Morley's ideas were the stuff of "cocktail party conversation" rather than science has become famous in the annals of the peer-review publication process. In a recent book the author says "scientists . . . generally give Morley his due, however, when they refer to the 'Vine-Matthews-Morley hypothesis.'" Just for fun, you might want to Google "seafloor spreading" and see how often you find the name Morley mentioned.

3. People at other places such as Rhode Island, Washington, Oregon, and Texas might argue about the Big Four terminology, and to be completely honest perhaps it should be called the Big Three and a Half.

was working on sponge fishery problems for the Colonial Office in the Bahamas Islands was introduced to the UM president at a cocktail party. This was even before Morley's work was suitable for such conversations, but the cocktails were good, and as they chatted, Dr. F. G. Walton Smith casually threw out the proposition that Miami was ideally situated to become the country's first tropical marine research center. He was an energetic and sparkling conversationalist, especially when suitably lubricated, and as he talked, he convinced first himself and then President Bowman Ashe, who immediately invited him to leave the Colonial Office and join the university as an assistant professor in the Department of Zoology with an eye towards organizing a marine laboratory, although at the time the University could provide neither space nor financial support.

Walton, as he was known, envisaged a horde of supporters to be garnered from the international sports fishermen and yachtsmen whom he knew from their regular visits to the Bahamas, and he plunged forward. He established the International Oceanographic Foundation as a base for that support, and on February 1, 1943, the University Board of Trustees formally established the UM Marine Laboratory.

Walton found an unused boathouse tied up on Miami Beach, raised enough money to rent it, and the Marine Lab was in business. Within a year, he had acquired a Navy contract for a study of marine fouling and another for the effects of tropical marine conditions on construction materials. When World War Two ended, Dade County (where Miami was located) built a causeway to two islands lying offshore in Biscayne Bay, Virginia Key and Key Biscayne. Among the plans was one for an aquarium, and Walton suggested that the Marine Laboratory operate it in exchange for land for a waterfront campus and a percentage of the admissions charge.

During the war the Navy had come to realize the value of science, and as the fighting war evolved into the cold war it began to pour money into oceanographic research. In 1958 Sputnik went up, and so did the country's research budgets. The National Academy of Sciences recommended that for economic and military survival the United States should inaugurate a long-range oceanographic agenda, and in 1961 the U.S. Navy laid out an ambitious $900 million, ten-year program. With the new onshore facilities and a small but fully equipped research vessel, the UM Marine Laboratory was in position to play a role in this burst of federally sponsored oceanographic research.

By 1963, when Harry Hess's idea about seafloor spreading had grown from poetry into theory, the Marine Lab had grown from a biological boathouse with two faculty members into a thriving research institution. In January of 1957, Cesare Emiliani had come from the University of Chicago, where he had pioneered oxygen isotopic studies on Foraminifera, microscopic organisms found in the oceanic sediments, which held important clues to the Pleistocene ice ages. The data indicated a rapid rising and falling of the isotopic abundances, which he interpreted as reflecting a similarly rapid variation in global temperatures. His results and interpretation, combined with the carbon isotopic record in the deep-sea sediments, enabled us for the first time to understand the conditions in past climates.

For over a hundred years we had been aware that several times in earth history much of the currently inhabited northern hemisphere had been covered by miles-thick glaciers of ice and snow. We had thought that these ice ages were three or four in number, and represented unusual conditions. But Emiliani's work showed that there had been many such icy periods, and in fact that the earth's climate rolled along on a roller coaster rather than on a level railroad track, varying rapidly and continuously from periods of intense warmth to periods of intense cold—and both these conditions are disastrous to contemplate. Our society obviously could not survive an ice age, with most centers of habitation buried under mountains of ice several miles thick. But the warm periods are equally disastrous, for the melting of the ice caps has periodically released enough water into the oceans to raise sea level hundreds of feet—and nearly all our centers of civilization are closer to sea level than that.

For the past ten thousand years we have been in an interglacial period and the climate has been reasonably steady. Emiliani's work showed that this constancy was an illusion, that the earth's climate varied continually with multiple episodes of ice ages. This led to our understanding that it is balanced on a razor's edge and ultimately to our realization that global warming is a real possibility.

Aside from his science, Emiliani was a unique individual. He spent the war years of the 1940s as a soldier in Mussolini's army, but by 1950—after obtaining a doctorate in Bologna and another one in Chicago—he was a postdoctoral student in Harold Urey's lab in that city, one of Urey's Three Musketeers, along with Gerry Wasserburg and Harmon Craig (who referred to themselves as the Geochemistry Mafia). Cesare "was a true Renaissance scientist," as Craig wrote, "at home in classical literature, fluent in many languages, and a dedicated opponent of dogma and mental rigidity wherever he found it."

Though he was one of the seminal figures in isotope geochemistry, my own favorite publications of his are a bit different. Toward the end of his life he was concerned with calendar reform, and in his final paper he showed that "even the Pope is confused," and corrected His Holiness's incorrect date for the beginning of the Great Jubilee of the Third Millennium. Thirty years earlier he wrote the shortest paper in the history of scientific literature. It began when a graduate student came to his office and said he had found an error in a paper Emiliani and several coauthors had recently published. Now the usual procedure, what the student expected, would have been for Cesare to send a correction to the journal. Instead, Cesare insisted that the student write a new paper, correcting the mistake, which the *Journal of Geology* subsequently published. In it, the student explained that in Cesare's original paper "the calculation of accumulation in grams per square centimeter was made on an average dry bulk density . . . [but] . . . the per cent water content by dry weight was misapplied to the equation..." In the acknowledgements, he wrote that he was grateful to Cesare "for encouraging the writing of this paper so that he could write the reply printed below." Then followed Cesare's Reply (anticipating Mr. Lewis), reprinted here in its entirety:

> "Absolute Dating of Deep-Sea Cores by the Pa^{231}/Th^{230} Method and Accumulation Rates": A Reply
>
> Oh, well, nobody is perfect.

When I came to Miami I attended my first faculty meeting with trepidation. We had held such meetings every Wednesday at Cornell, and every Wednesday evening I would finally return home with a migraine from the stress of sitting politely with my cup of tea while the chairman and the faculty, all dressed somberly in suits and ties, droned on and on and on. But at that first meeting in Miami, Cesare bounded into the room wearing shorts and sandals and carrying a large carton from which he withdrew cans of beer and threw them around the room to all of us, and I thought I had died and gone to heaven.

In 1995, a couple of years after retiring from the University of Miami and just a few weeks after the publication of his final paper, Cesare did that literally, pitching forward into his pizza with a beer in his hand, exactly the way he would have liked to go[4].

But now we're back in 1957, when Dr. Smith told Emiliani he wanted to increase the Lab's expertise in the physical sciences, and Emiliani suggested inviting Fritz Koczy, an Austrian scientist whose area of research covered the entire field of oceanography and who was working at the oceanographic institute in Göteborg, Sweden. The two of them established a Division of Physical Science and began hiring outstanding people who brought in tons of federal research money. And then, in 1965, the University administrators

4. Or so I was told. I hope it's true.

established a School of Environmental and Planetary Studies, brought in Fred Singer as Dean, and told the marine triumvirate—Walton, Fritz, and Cesare—that they were to be bundled into it.

I don't think so, they said. Cesare was the flamboyant one, and he wanted to march into the president's office and tell them all to fuck off. But Fritz and Walton were more skilled at university politics, and they reined him in. Instead of protesting, they simply worked behind the scenes to veto everything Singer tried to do; for example, hiring me. Finally, by the time of the 1966 Spring AGU meeting in Washington, the UM administration conceded defeat, fired Singer, and disestablished SEPS. The Marine Laboratory was once more free, independent and unfettered, and Cesare called out joyously to me on the steps of the Sheraton Hilton, "We got rid of Singer," he said. "Now we want to hire you. The oceans are spreading!"

The way science usually works is that someone makes an observation and someone else (usually) puts together a theory to explain it and then some other people (usually) work out a way to test the theory. In this case Wegener, along with a lot of other people, had made the observations: that the continents seemed to fit together like a jigsaw puzzle, that a few of the same fossils were found on different continents, and that ancient simultaneous glaciations on different continents could not be explained. Wegener alone had the courage to propose a theory to explain all this, but it had the fatal flaw that the continents couldn't drift around through solid rock. Hess then proposed that they actually were sitting on top of the seafloor, which spread out from the ridges and sank into the trenches, and when Vine and Mathews (and Morley, remember) observed magnetic striping patterns identical on each side of the ridges, it suddenly all made sense.

It may have become a reasonable theory rather than "geopoetry," but it wasn't yet proven. Cesare and Fritz realized that if

The author, fiddling with the mass spectrometer extraction line, 1959. (Photo by Leila Fisher)

Grenville Turner, 1963. (G. Turner)

John Reynolds, ca. 1973. (University of California, Berkeley)

Ray Davis and John Bahcall at the Homestake Mine, 1964. (Photo by Don Harmer. Courtesy of Neta Bahcall/The Shelby White Leon Levy Archives Center, Institute for Advanced Study, Princeton, N.J.)

Tommy Gold, ca. 1960s, as I remember him. (Cornell University)

Cesare Emiliani, ca. 1960s. (H. Wanless)

Hess was right, if the ocean floor really was being produced at the mid-ocean ridges and spreading apart, then the rocks on the ocean floor next to the ridges had just erupted and would be young, and as they spread further away they would grow successively older. All you had to do to test this theory that was sweeping across the world was to measure the ages of the ocean floor at various distances from the ridge. And since the ocean floor was composed of volcanic rocks, the potassium/argon method was ideal. What they had to do to take the lead in this new field was to hire someone who could do K/Ar dating.

Enter: me. From my visit the previous fall and our discussions they knew I had been measuring the K/Ar ages of meteorites. They didn't know that the technique I had been using was not the right one to use on the oceanic rocks. The iron meteorites had so little potassium and argon in them that I had had to use neutron activation, which is a technique with fantastic sensitivity (i.e., it can measure small amounts of stuff) but with less precision than mass spectrometry. To measure accurate ages on the oceanic rocks, mass spectrometry would have to be used. This ignorance on their part was balanced by another ignorance: they didn't know that I knew how to do mass spectrometry from my time at Brookhaven. The ignorances canceled out: it turns out that sometimes two wrongs do make a right.

Cesare called Gerry Wasserburg to ask about me, and Gerry—despite (or perhaps because of) his contretemps with me—gave a glowing recommendation, and so I bounced down the steps of the Washington Statler Hilton and flew home to tell my family we were going to Miami instead of Oregon.[5]

To measure the ages of the rocks on the spreading seafloor.

5. We arrived there that August. My previous visit had been in November, and it had seemed like paradise after the bitter cold of Ithaca. Now we stepped out of the airport into an incredibly oppressive heat garnished with 100% humidity, and I thought "My God! What have I done? No one could live in this." Marshall, my three-year-old son, clasped my hand and looked around in wonder. I had told him I was taking him to this great place, Miami. Now he asked, "Daddy, is *this* your Ami?"

THIRTEEN

The Argon Surprise

This is disgusting!

—Little old lady

THE FIRST THING I DID IN MIAMI WAS TO WRITE a proposal to the National Science Foundation for a mass spectrometer, in order to test Hess's idea of a spreading seafloor. Funding was not a problem in those halcyon and bygone days of yore. Once, I remember, Cesare came trotting down the hall calling out that it was the end of the fiscal year and the NSF was on the phone; they were calling to say they had two hundred thousand dollars left over from the budget, and did anyone want it? No one did, we all had enough money.

Lordy, lordy. (Loud sigh.)

And so the money for the mass spectrometer came through, but not before summer, and I was not about to spend July and August in the Miami furnace. Instead, I arranged to go up to the State University of New York at Stony Brook, where Ollie Schaeffer had become head of a new earth sciences department, to use his mass spectrometer to measure the ages on a suite of rocks brought back by one of my new friends at Miami, Enrico Bonatti, a marine geologist who had just returned from a research cruise with ocean floor samples that were perfect for testing the spreading seafloor hypothesis. He had dredged up basalts from the flanks of the East Pacific Rise and a half dozen other samples at various distances from it. So we should see young ages on the ridge rocks, and a spectrum of increasingly older ages as we moved outwards. Basalts are good material for normal potassium-argon dating, and those on the seafloor, we thought, should be even better. The basis of K/Ar dating is that you have a magma region somewhere inside the earth, with potassium continually decaying to argon. When the magma erupts, throwing out molten basaltic rocks, all the argon previously produced will bubble out and be lost to the atmosphere; as the lava cools into basaltic rocks, they will have potassium in them, but no argon, effectively setting the dating clock to zero. As time passes, more argon is produced, but this posteruptive argon is now trapped inside solid rock. The potassium content therefore is continuously being depleted, while the argon content is increasing; the ratio of potassium to argon in the rocks is all that is needed to determine the time when they cooled and argon was retained, and so a measurement of both potassium and argon content will effectively determine the age of the basaltic sample.

Enrico's deep-sea rocks would be perfect samples for dating. As mentioned previously, it's hard to be sure that rocks are "airtight"; some fraction of the argon might leak out over millions of years, providing an age younger than the true one. But these oceanic rocks erupted onto the seafloor, where the overlying pressure of the water and the cold ambient temperature would be sure to inhibit any diffusion loss.

And after six years of isolation at Cornell, I suddenly had collaborators coming out of my ears: Oiva Joensuu was a consummate analytical chemist who would do the potassium measurements at Miami, John Funkhouser at Stony Brook had Ollie's mass spec in working order, Enrico had brought back the samples, and all I had to do were the routine measurements.[1]

1. As at many of the new oceanographic centers, the faculty at Miami was international. Kurt Bostrom was a Swedish geochemist, Walton Smith was English, Koczy was Austrian, Cesare and Enrico Bonatti were Italian, Göte Östlund was Swedish, and Oiva Joensuu was Finnish. In fact, the only American I worked with there was Joe Prospero, from Pottstown, Pennsylvania.

I settled my family in at a slimy Connecticut beach on the polluted Long Island Sound shore, and I took the Port Jefferson ferry to spend Mondays through Thursdays at Stony Brook. The family loved it; my wife Leila had grown up with the Connecticut beaches, and the kids didn't know any better. Meanwhile, I had the wild Atlantic surf and a fully functioning mass spec for four days a week and then long weekends with my family. Life can be wonderful.

Except...

The measurements didn't turn out exactly as expected. After a couple of weeks calibrating the instrument and running test samples, John Funkhouser and I plunked in our first real sample, a glassy basalt[2] from right on the EPR (East Pacific Rise). This was to be our baseline sample: if we knew anything at all about the mid-ocean ridges, it was that they were volcanically active, and so any basalts found there would be geologically young—thousands of years old perhaps, at most maybe hundreds of thousands—while if the rest of the ocean floor was made of material spreading from the ridges, it would get successively older, reaching ages of many millions of years.

Right. Except that this first sample, from the very crest of the EPR, didn't behave as expected. We plunked it into our glass vacuum line, melted it, cleaned up the gases, and pumped them into the mass spec—and found tons of radiogenic argon.[3] Combining this

2. Volcanic basalts on the ocean floor erupt into temperatures barely above zero, and the molten lava freezes so quickly that mineral crystals don't have time to form. The result is a shiny black glassy skin, millimeters to centimeters thick.

3. The earth's atmosphere consists of 1% argon, most of it argon-40 from the decay of crustal potassium-40, along with remnants of primordial argon-36 and a little bit of argon-38. This atmospheric argon contaminates every measurement, since no experimental technique can get rid of it completely, but it is recognizable by a constant 40/36 ratio of 295.5. So to measure the radiogenic argon-40, in order to calculate a K/Ar age, one measures the total amounts of 40 and 36, multiplies the 36 by 295.5, and subtracts that from the measured argon-40. The remainder should be pure radiogenic argon.

with the potassium values Oiva Joensuu had already measured for us, we quickly calculated an age for this rock of 500 million years.

"Yippee," I shouted, and called Miami to tell Enrico the good news. (Nothing is more fun in science than finding evidence to prove that someone, particularly an "authority"—in this case, Harry Hess—is totally wrong.) But no, Enrico said, not possible. A geologist, he knew what I did not, that no basalt could exist for hundreds of millions of years in salt water without showing major signs of alteration, and we had purposely picked this particular rock because it was fresh, clean, unaltered. Plus, undersea photos of the dredging site showed the rocks on the crest were bare, uncovered by the sediment that is constantly falling to the seafloor. No way it was this old, impossible for it to have been sitting there for half a billion years.

I had a sudden epiphany: I was cursed. I had spent more than five years chasing the spectre of iron meteorites older than the solar system only to find finally that the old ages were a mirage, and now here I was again! What the hell was happening?

The first thing to do was to find out if this rock was unique, so we spent the next few weeks measuring the argon in all the rocks I had brought with me. It turned out that all the basalts with glassy rims had ages of hundreds of millions of years, with no relation to their place of origin, that is, they didn't increase with distance from the ridge. Some of the basalts had evidently cooled more slowly, with no glass showing, and these all had much lower ages.

So, first of all, it was clear that the basic criterion for a valid K/Ar age—that all the pre-eruption argon was degassed upon eruption, effectively setting the clock to zero—was violated in the glassy basalts. They had frozen immediately upon hitting the cold ocean waters, and whatever argon they had before erupting was frozen into them. Their ages were meaninglessly high.

The crystalline, or nonglassy basalts, presented another problem. They had risen to the surface slowly enough for crystals to form, pushing out the magmatic argon as they did. But we couldn't tell if they were leak-tight for the posteruption argon; that is, their measured ages might be too young. We weren't able to find any samples that would give trustworthy K/Ar ages.

By the end of the summer I was beginning to get depressed. It was clear by August, when we packed up and left to head back to Miami, that the basic reason they had hired me had fallen apart. So what was I going to do with the rest of my life? I was still doing some work on meteorite ages, but that was dying another kind of death. The oceanic rock work had expired with the first age we measured, an obviously wrong five hundred million years, and the meteorite work had just gotten boring. We had hoped that measuring such ages would lead to interesting theories of solar system formation, but it really didn't. At least it hadn't so far, and I was beginning to think that it never would.[4]

So there I was, flat on my back and without a sign to point out a new direction. I was thinking I might quit science entirely and become an actor, when Paul Gast walked in the door.

I was serious about becoming an actor. I had played leads at the semi-professional Oak Ridge Playhouse and with a community group called the Barnes Players in Ithaca. When we moved to Miami, a professional repertory theater was having its first season, playing in the historic Coconut Grove Playhouse, and I thought they were really good. In the spring I auditioned for them, and landed the leading role in Lorraine Hansberry's *The Sign in Sidney Brustein's Window*. At fifty dollars a week, I thought I was a real actor.

For the first time, I felt I knew what I was doing in a tremendous role. I was onstage for the entire play, which was voted the

4. As it turned out, I was right. The meteoritic community had a lot of fun and spent a lot of money arguing about the reliability of different experimental methods and the meaning of the measured ages, but nothing of real importance has ever come out of it. What did turn out to be important was the measurement of short-lived radioactivities, beginning with xenon. We'll get to that.

best play of the year in Miami. (Okay, it was only Miami, but that vote put us ahead of a whole season of professional road shows out of New York.) I also began to write plays, and the directors of the Miami Actors Company, as we were called, decided to produce my newest one in hopes of raising money for a New York run.

The play was called *The Courtesy Not to Bleed,* and it was a two-character romantic comedy featuring a black man and a white woman. This was in 1967, in Miami, which in some respects was New York South, but was still located geographically and culturally in the Deep South. There were signs on the drinking fountains in the train station reading "White" and "Colored," the schools were still segregated, and I made a fool of myself by loudly complaining, at the tennis club I joined, about the sign in the pro shop which read "Please obey the all white rule." I was told, amid laughter, that it referred to the clothing we were supposed to wear. (Still, I never did see a black player there.)

So all in all, the play was maybe a little ahead of its time. We had to get police protection because of bomb threats, but that wasn't the real problem. The male lead was played by Ray Aranha, who later went on to Broadway, and he was very good. The actress he played with was not. I was seeing myself as a combination of Pinter and Ayckbourn, and I saw the play as having a lot of funny lines but the real impact was to be garnered by playing under the lines, with the actors bringing out a sensitive relationship aborning.

So rehearsals began, and by the end of the first week I was nudging the director: "They're not getting it, the play is flat."

"Not to worry," she said, "they're still getting their lines down. Early days yet."

By the end of the third week I was getting nervous. "There's nothing there," I complained. "They're just going for the laughs. There's no meaning."

She laughed and patted me on the arm. "All you authors are alike. Give them time, they're pros. They'll get it right."

By the sixth and final week I was screaming at her, and her reply was suddenly different. I still remember her final words. "That's as good as it's going to get. I can't get anything else out of them."

What had happened was that the director had searched Miami for the best actress, without asking her what she thought about black men. It turned out she hated and feared them. She took the part because it was a leading role, and no actress below the Meryl Streep class turns down leading roles, but she just couldn't get over her antipathy towards Ray.

And it showed. Ray picked up on it, and we had a disaster brewing. It was supposed to be a romantic comedy—and the two actors hated each other. It reached a climax in dress rehearsal. The first act builds to a scene in which she's supposed to slap him, he slaps back, and as he does she catches his arm and flips him over her shoulder. But at the final dress rehearsal, when she slapped him, she hit him hard and it was the legendary straw and the camel's back: Ray swung back and smashed his open palm against her face and knocked her clear off the stage.

She dissolved in hysterics and fled to her dressing room. Ray sat down on stage and stared at the floor, muttering "Bitch, bitch, bitch," over and over again.

We opened the next night.

Catastrophe.

The two leads glowered at each other, so that the charming words of awakening love made no sense at all. And when finally the black man took the white woman in his arms and kissed her, a little old lady with blue hair sitting front row center stood up and announced to the audience, "This is disgusting!" She turned back to the stunned actors onstage, frozen in their obligatory embrace, and repeated the one word: "Disgusting!" Then she stalked out of the theatre.

Needless to say, we were not a hit. In fact we were so not a hit that our repertory group tossed in the towel; the directors had been counting on ticket sales of the play replenishing a bankrupt treasury. So there I was, a playwright without a play or an acting company and a scientist with not the faintest idea of what to do with the mass spectrometer that the NSF had just given me a quarter of a million dollars for.

It was the next day that Paul Gast walked in the door. He was one of the country's leading geochemists, working out of the Lamont Geological Observatory (now Lamont-Doherty) of Columbia University. His early death in 1973 was a sad blow to earth science. This day, however, he was in good health and high spirits. He was returning from a research cruise that had docked in South America, and he had stopped for a visit in Miami on his way home. I had never met him before, but he strode into my office and said, without preamble, "You are one lucky bastard."

"Huh?" was all I could say.

"I sat on your NSF panel," he explained. So he knew I was getting a mass spectrometer from them.

I shook my head. "Not so lucky. It turns out we can't measure the age of the—"

"That doesn't matter!" he practically shouted. It turned out that he was also one of the referees for the paper we had sent in to *Science* reporting on the problem of excess argon in the oceanic rocks. "Doesn't matter at all," he repeated. "There'll be other ways of measuring ages,[5] but you've got something unique!"

"I do?"

He laughed. "What you've got here is a deep earth probe. You can look right into the mantle!"

5. He was a bit optimistic about that. Twenty years later, in 1987, I attended a workshop at Northwestern University convened to look into possible methods of dating oceanic rocks. There was still no way of doing it. Even today there is no generally applicable method, although in some circumstances reliable ages can be determined.

With that he waved goodbye, he had a plane to catch, and woof! He was gone, and I was left there wondering, in effect, who was that masked man?

Slowly I began to realize what he was talking about. We had, at the time, no reliable information on the inside of the earth. If you think of the earth as an apple, we had no samples from deeper than the skin. All the rest was unknown territory. But these oceanic rocks we had measured had trapped and retained their argon from somewhere much deeper than that, and so they were providing us with unique information. I sat there and began to think of argon not as a dating tool but as a relic of the primordial earth. What did we know about argon, and all the other noble gases, in terms of how the earth formed and evolved?

The basic information they gave us is actually incorporated in one of their names—the rare gases. They are not at all rare in the universe or, more particularly, in the sun.[6] Figure 3a shows the abundance of the elements in the earth; figure 3b shows the same plot for the sun. The very fact that the noble gases are so rare on earth told us that the accreting earth was essentially incapable of retaining an atmosphere and that therefore the earth's atmosphere must have resulted from a later degassing of the solid earth.

What I could do now, following Paul Gast's suggestion, was to go further, using the argon isotopic composition measured in the oceanic rocks. At the time, 1967, the only question was: when and how did this degassing occur? There were two endpoint possibilities.

Option One: Suppose the earth formed as a collection of chondrites, as Urey suggested. Remember, the nonvolatile abundances

6. "More particularly" because the earth formed, somehow, as a remnant of solar formation. This much we know, but not much else.

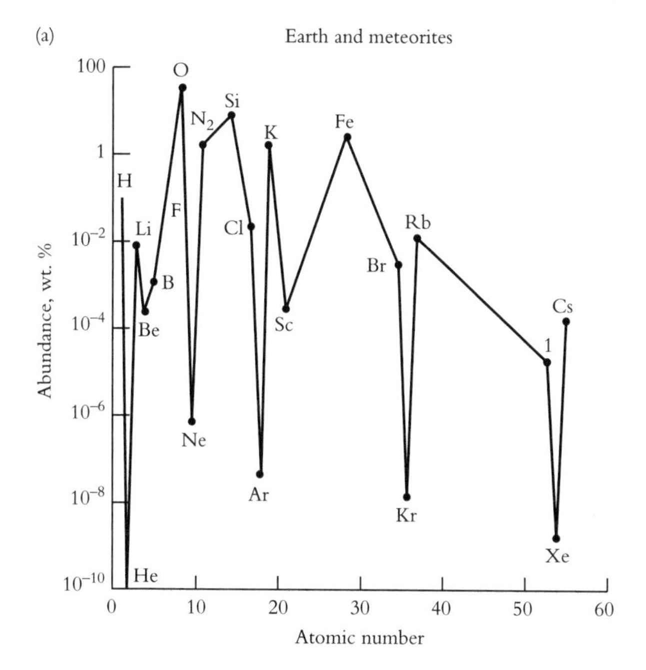

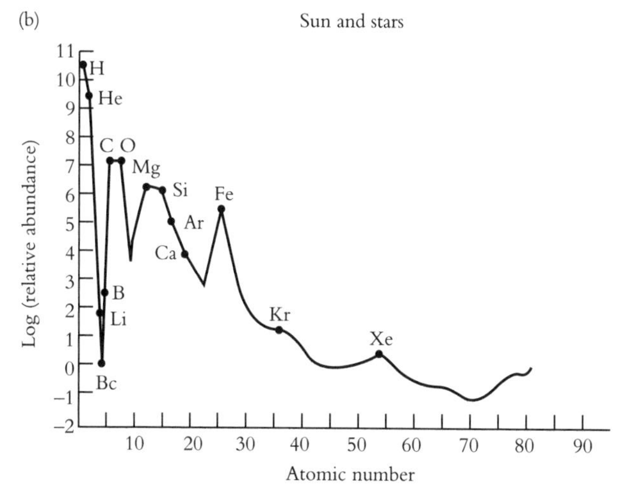

FIGURE 3A, 3B. Rare Gases: Earth, Sun and Stars

of the chondrites and the earth are similar, nearly identical. The main difference is that in the chondrites the iron is distributed throughout, while in the earth it's concentrated in the core. So the idea is that "soon" after a chondritic earth was formed it melted, either from the trapped heat of the accumulated chondrites smashing together or from radioactive heat generated by uranium, thorium, potassium, and possibly other short-lived radioisotopes that are no longer present. During this melting phase, the iron sank and formed the core, while the volatiles in general and the noble gases in particular were vented from the interior and formed the atmosphere.[7]

So when did this Big Burp occur? We didn't know. The most we could say is that if it was caused by the trapped heat, which would quickly dissipate, it had to be soon after or even during earth formation, but if it was radioactive heat, there would be a roughly billion-year delay as the heat built up. (And of course it could be a combination of the two, possibly with now extinct short-lived radioactivities thrown in.) You can see that the theories are badly constrained.

And now consider Option Two: the earth formed its core first, accumulating iron meteorites, then added a veneer of chondrites. In this model there is no need for planet-wide melting, thus no Big Burp. The atmosphere would evolve slowly over time by volcanic degassing.

I began to get excited. Paul Gast was right, and I was stupid not to have thought of it myself. The basic point is that the ratio of argon-40 to argon-36 in our atmosphere is 295, but nuclear theory tells us quite definitely that the isotopes were formed with a ratio much less than one. Clearly the ratio has increased over time due to the decay of potassium-40 (to argon-40). Now consider the two end possibilities of a Big Burp formation of the atmosphere: it occurred either early in earth's history, or just yesterday. (Bear with me.)

7. The lightest gases, hydrogen and helium in particular, would be lost to space.

If it occurred yesterday, the 40/36 ratio of the gases that came out would be the same as those left behind, that is, the ratios in the atmosphere and in the mantle today would be identical. But our measurements showed that the trapped (mantle) argon has a ratio more than an order of magnitude higher than atmospheric, so right away we can say that the atmosphere did not form yesterday!

All right, so we already knew that. But the point is that the later the atmosphere formed, the closer would be the atmospheric and mantle 40/36 ratios. And now I had a method of measuring the mantle ratio! I could determine how the atmosphere formed, and through that constrain how the earth formed! Late, continuous degassing would rule out the Big Burp and therefore core formation by planetary melting; Option Two would be proven. Or a wide variation between mantle and atmospheric argon would prove an early Big Burp and consequently earth formation by chondritic accumulation.

Of course, there were still a few details to be worked out...

FOURTEEN

Primordial Helium and Argon and the Evolution of the Earth

Dancing in the dark, 'til the tune ends
We're dancing in the dark . . .

—H. Dietz

. . . And a few surprises in store. But surprises don't surprise us; they're expected. We don't know what they'll be, but we know they're lurking somewhere out there in the vast unknown of our barely investigated universe: "Seek and ye shall find, but seek not to find that for which you seek". . . or you'll miss the important stuff.

Finally my new mass spectrometer showed up, a custom-made machine put together by Nuclide Analysis Associates, a group operating out of Penn State, consisting of one professor and—most importantly—a fine technician who showed up with a parcel of crates and proceeded to put it together. In only a few weeks, we had a beam of ionized noble gas ions and began to try to make the instrument work.

We had, at that time, with that machine, no means of determining the mass of the ions being sent down the beam to the detector. The heavier the mass, the stronger would have to be the magnet setting in order to bend it out of a straight path and deflect it into the detector, so—as when I started at Brookhaven—it made sense to start by trying to find the lightest element, helium. In particular, I went looking for the ^{4}He isotope, both because it was radiogenic (formed from the decay of uranium and thorium) and because it was by far the most abundant (and thus most easily measured) isotope of helium.

I started by melting one of my ocean rocks, cleaning up the gases, and searching among them for the nearby mass 2 peak, which was unwanted but always present due to the ubiquitous hydrogen molecule, and soon found it. Then I scanned upwards and quickly found the helium-4 peak. To my surprise, there was a tiny little bump in the scan between the two, at mass 3—unexpected, like a ghost that shouldn't be there.

It couldn't be hydrogen-3, tritium, because that's radioactive and there isn't any method of producing it in the oceans—well, there was still tritium wandering around the ocean waters from the nuclear bomb testing of the 1950s, but it was extremely unlikely to have found its way into the solid rock of the ocean bottom.

It could be the light isotope of helium, helium-3, but the amount formed in the radioactive decay of uranium and thorium—the source of the helium-4—was too small; that is, the peak I was looking at was too big, compared to the ^{4}He peak, to be radiogenic ^{3}He.

It was probably the HD molecule, formed with one normal hydrogen (mass = 1) atom and one of the rare, heavier (mass = 2) hydrogen atoms. So that was that, and I went home and to sleep—and in that sleep remembered something one of our Ithaca friends, Ian Axford, had been talking about at a party a couple of years ago. It had been a wintry night with light snows when the people came, but at three in the morning, when it broke up, we found that everyone was snowed in, and Triple A spent a while towing the cars out of our driveway (a not infrequent event in Ithaca). Ian had been working on an idea about the origin of helium-3 in the atmosphere.

First thing in the morning I went to the library and found that he had published his idea just the previous year. He argued that the (vanishingly small) amount of ^{3}He in the earth's atmosphere was being continuously produced in the solar wind, and was lost just as quickly back into space. Interestingly, at just about the same time, Harmon Craig and his student Brian Clarke had argued instead that the ^{3}He was primordial—that the earth was formed with a chondritic ^{3}He abundance and was not completely degassed, so that the original amount of ^{3}He (and, of course, other primordial gases not yet detected) was continuously leaking out through the ocean floor.

Both Ian and Harmon were extremely bright scientists,[1] but Ian was so much more likable that I found myself hoping he was right. More importantly, I thought I had the data to find out. The mass 3 peak in my data was HD, I thought, but if I assumed it was ^{3}He it provided an upper limit to the amount present; in other words, if there were more ^{3}He in the sample, the peak would be higher. And this upper limit was enough to prove Ian right: the helium-3 in the atmosphere was from the solar wind rather than earth's interior. I could state that if the earth at one time had a chondritic ^{3}He abundance, it had long since degassed thoroughly. Happily, I published this result.

Unhappily, I was wrong.

Not in what I said in that paper, but in what I thought, which was to confuse the word "thoroughly" with "completely." I assumed that any trace of primordial gases had long since disappeared from the earth's mantle, and that the future of this line of research lay with the radiogenic gases ^{4}He and ^{40}Ar. These, being continually

1. Axford left Cornell at about this time to own a motel in La Jolla and do research at Germany's Max-Planck-Institut für Aeronomie. He later was director of that Institute and, as a native-born New Zealander, was knighted in 1996. He now resides in New Zealand, where he heads that country's research in radio astronomy. Harmon Craig spent his entire career at Scripps Institute, and was a potent (and stridently vocal) force in geochemistry until his death in 2003.

produced and volcanically degassed, should trace the development of our atmosphere. So when Harmon Craig later found evidence of primordial helium-3 in ocean waters and Igor Tolstikhin (in Russia) found it in well gases, I thought it was mildly interesting; if confirmed, it meant that the earth hadn't completely lost its original gases, but so what?

Meanwhile I was concentrating on argon, and getting a bit worried about whether there was any homogeneity to the internal argon ratios. My first data were showing a large variability from rock to rock, and even within different portions of the same rock. The outside (the atmosphere) is homogeneous, but if the inside of the earth is not, how can the two be compared? The interpretation was getting very complicated, perhaps too much so for mortals to crack.

Or so I thought. But while I was stewing in Miami, a new protagonist entered the lists. Minoru Ozima, at the University of Tokyo, began to beget a whole new generation of talented researchers. His first publications on the subject, in 1972 and '73, not only presented more argon isotopic data but dove right on into interpreting them in terms of atmospheric evolution. I thought he was jumping the gun, but then two things happened. In 1975, Harmon Craig found ^{3}He in mid-ocean-ridge basalt (MORB), and suddenly everybody was finding it and it quickly assumed great importance,[2] as a growing number of studies showed that the ^{3}He/^{4}He ratio was quite constant in all MORB samples, at nearly ten times the atmospheric ratio of 1.39×10^{-6}. Further studies showed that the helium ratio is even higher (and more variable) for oceanic islands. At first, this difference between the two regions was widely construed to be the most important information to come out of oceanic noble gas studies, implying

2. The amounts found in MORB clearly indicated that my mass-3 peak indeed contained ^{3}He as well as HD. If I had been a better experimentalist, I could have tuned the spectrometer properly, resolved the peak into two separate ones, and made this important discovery. As it was, by reporting only an upper limit, I missed it.

that the mantle is composed of at least two distinct regions, a well-degassed convecting mantle underlying the crust and giving rise to the MORB, and a (presumably) deeper, less degassed region feeding the oceanic islands. But as more data accumulated, the interpretation became more convoluted, so that today ... well ...

In Miami, meanwhile, I was ignoring the helium data and getting excited when another Japanese group used a new method of graphing their results which showed clearly that the argon isotopic variability in our data was due to varying amounts of atmospheric contamination.[3] This opened the gates, and Minoru Ozima and I quickly tried to interpret the data on the assumption that the highest measured $^{40}Ar/^{36}Ar$ ratio in MORB was a lower limit to the mantle value, with lower measured ratios being due to atmospheric contamination (the atmospheric ratio is 295.5, while we were getting MORB values reaching into the tens of thousands). Things began hotting up, and in 1977 Ozima and John Reynolds (see chapter 14) convened a joint U.S.-Japanese conference to discuss the results.

Discuss: that's a polite scientific word meaning argue vociferously. The object of all this, remember, was to understand the evolution of the mantle and atmosphere.[4] In particular:

1. Did the atmosphere originate primarily in one early burp or in later continuous volcanic degassing? And
2. Is the mantle homogenous, everywhere similar to the degassed (depleted) mantle regions feeding the MORB, implying whole-mantle convection, or is there somewhere a separate, largely undegassed and nonconvecting mantle region?

At the time of the conference, we had hoped that ^{40}Ar studies would answer these questions, but it soon became apparent that

3. This was not a problem for the helium work because helium is so incredibly rare in the atmosphere, while argon is present at 1%.

4. The earth's core is beyond the reach of any noble gas data, and the crust is too heterogeneous for any simple argument.

the ^{3}He (and later neon) data were hugely important. So, to skip the details and bring you up to date, I now present a bunch of "on the other hand" arguments which you may wish to skip to preserve your sanity. On the other hand, this is the way science progresses, isn't it? At first you know nothing, then you suppose something, and then (if you're lucky) you learn—but not what you were supposing.

So, then.[5] We don't know the whole Earth K/U ratio. But if we suppose—as some arguments suggest we might—that it is much lower than the MORB value, then it follows that the ratio of ^{40}Ar in the MORB compared to the atmosphere implies convection and mixing throughout the whole mantle, and if this is so, then the depleted mantle that serves as the MORB source is in fact the whole mantle; in other words, the models based on the helium-3 differences, which invoke a separate gas-rich mantle reservoir, are wrong.

On the other hand, argon isotopic studies of basalts from ocean islands (OIB, or Oceanic Island Basalts) show much lower 40/36 ratios than do the MORB, inspiring those who found these data to agree with the helium-3 interpretation which supposes that the OIB mantle source is largely undegassed and therefore contains primordial argon (which would have more ^{36}Ar and thus a lower 40/36 ratio), and thus constitutes a separate mantle region. But, on the other hand, the lower argon ratios are probably due merely to atmospheric contamination (the atmospheric ratio, remember is only 295.5).

To continue with the supposed lower K/U ratio, comparing the ^{4}He to the ^{40}Ar in MORB suggests that later continuous volcanic degassing is responsible for most of the atmosphere. On the other hand, Ozima argued that the high ^{40}Ar/^{36}Ar found in MORB could be explained only by early catastrophic degassing, but I showed that taking potassium mantle/crust fluxes into

5. See Notes for references to the following.

consideration made them perfectly compatible with later continuous degassing. On the other hand, a recent review found (without going into much detail) that Ozima's argument was the more convincing.

And finally, consider the nonradiogenic isotopes. Neon has three and xenon has seven, and several recent studies have found that the isotopic ratios in both elements are different in the mantle than in the atmosphere, and are almost certainly due to mass fractionation. Such fractionation can take place, for example, when gases are lost to space, with the lighter ones going at a faster rate, or similarly by diffusive processes within the earth as the gases pass from one reservoir to another. This raises another level of difficulty in interpretation, and in fact the review previously discussed, although stating at several points that the data require the early degassing model, goes on to argue that this fractionation "renders invalid all calculations of degassing of the solid Earth based upon matching upper-mantle isotope compositions by simple degassing into the atmosphere. Rather, the relationship between mantle noble gas compositions and the degassing of the atmosphere is more complex" and not yet understood.

In fact, it's possible that the atmospheric and mantle gases have nothing to do with each other. It's been suggested that the atmosphere arose not from degassing at all but from a veneer laid down on an already formed earth by outside sources such as comets. A detailed description of possible origins and their relation to earth formation and evolution models, together with pertinent references, can be found in Ozima and Podosek's fine monograph, *Noble Gas Geochemistry*. Since no positive conclusion can yet be drawn, it doesn't seem worthwhile to spend more time on the subject in a book such as this, except to note Ozima and Podosek's summary, "there is still no satisfactory theory on the origin of terrestrial noble gases."

Well. Does this embarrass or discomfit us? Not at all, although it is a bit depressing. As we said earlier, this is the way science progresses. It's a dark jungle, this universe of ours, unlit by even the faintest of moons, and we're staggering through it, trying to find our way to enlightenment, taking false steps, tripping over ourselves, doubling back and starting over—but the weird thing is that slowly—sometimes very slowly—we actually make some progress, and over the eons we begin to understand. As Einstein said, "The most incomprehensible fact about the universe is that it is comprehensible."

This is a faith backed up by the most incontrovertible fact of scientific history: little by little we have learned more and more. So some day, I have no doubt, we will understand the relations between the oceanic basalts, the atmosphere, and earth's evolution. In the meantime, however, we are still dancing in the dark.

And so were all these studies a tremendous waste of time, simply because at present we don't understand their meaning, don't understand how the atmosphere formed and how the mantle is structured? No, not at all, and perhaps in the second edition of this book (or maybe the nth), I'll explain it all.

Meanwhile, let's talk about something a bit more successful. Let's talk about xenon…

FIFTEEN

Xenology

Raffiniert ist der Herrgott, aber boshaft ist er nicht!

—A. Einstein

XENON IS UNIQUE AMONG THE NOBLE GASES IN that it has an isotope, ^{129}Xe, that is the fossil daughter of an extinct nuclide. Iodine-129, its precursor, decays to ^{129}Xe with a half-life of about sixteen million years, and since the earth is four and a half billion years old (and since all the elements on earth were created in stars before the earth accreted[1]), there is no ^{129}I on earth today; after the first hundred million years of earth's existence there would have been less than 2 percent left, after a billion years there would have been too little to measure, and by today we can safely say there is "none" left.

But now let's go back to the very creation of the solar system. We know that the elements that exist today were created earlier in stars and blown out into space, and somehow they accreted into the sun and planets. We know roughly how and in which types of stars the elements were created, but we still don't know the details of their synthesis, and we know even less about their

1. Not strictly true. Cosmic rays impact on us, creating, for example, carbon-14 and a few other isotopes, and we create new ones in our laboratories, and others such as ^{40}Ar are continually being created through radioactive decay, but none of this applies to iodine.

accretion into the sun and planets, and until the xenon studies we had absolutely no idea *when* they were created.

Suppose that the creation of the elements took place billions of years before solar system formation (after all, the universe is nearly ten billion years older than we are). Then all the ^{129}I would have decayed into xenon long before the sun and planets formed, the ^{129}Xe would have mixed with all the other xenon isotopes, and upon its incorporation into the solid particles of the solar system the xenon would be isotopically homogeneous. The sun, the earth, the meteorites, and the planets and moons would have incorporated differing amounts of xenon, according to their mode of formation and evolution, but they would all have the same mix of xenon isotopes (with perhaps some easily recognized mass fractionation).

But suppose instead that the elements were created just previous to solar system formation; that is, within a few half-lives of ^{129}I—say, less than a hundred million years. Then as the first solid mineral grains formed in the solar nebula, some iodine—in particular the ^{129}I isotope—would still exist and, according to the chemistry of the various grains, would be incorporated into them in greater or lesser amounts. This ^{129}I would subsequently decay to ^{129}Xe, which would be trapped within the solid mineral. If this scenario held true, the important consequence would be that somewhere in the solar system there would be grains that incorporated a large amount of iodine at the beginning, and today would show an excess of ^{129}Xe.

The geochemistry of iodine was largely unknown; it was an element of negligible interest, and so no one knew exactly where to look for the xenon excess. In fact, almost no one was interested in looking for it, because a similar search had already failed. The impetus for the search had come from Harold Urey, a Nobel-winning geochemist at the University of Chicago (and the scientific mentor of the *drei Wunderkinder* Gerry Wasserburg, Harmon Craig, and Cesare Emiliani). The time was the mid-'50s, and Urey was working out a theory of planetary formation that involved smaller planetesimals forming first, melting, and then aggregating into today's planets. The melting was a necessary but obstinate part of his theory; necessary to provide for the chemical differentiation he needed in order to account for today's planets, obstinate because his proposed planetesimals weren't large enough to melt.

The terrestrial planets melted, according to his theory, by retaining the heat generated by radioactive disintegration of long-lived potassium, uranium, and thorium isotopes (thus differentiating into core, mantle, crust, and atmosphere). But Urey's smaller planetesimals would have had a higher surface-to-volume ratio, and so would have radiated heat from their surfaces too fast to allow their temperatures to build up to the melting point. To get away from this difficulty, he proposed that they must have formed soon enough after element creation to allow for the existence of short-lived radioactive isotopes (those with half-lives of millions of years, instead of the billion-year half-lives of the "long-lived" isotopes) to provide additional heating.

Iodine is an element of low abundance, so Urey didn't expect that ^{129}I would provide enough heat to make a difference, but just the previous year—in 1954—the perfect isotope had been discovered: aluminum-26. Aluminum is one of the more abundant elements, and ^{26}Al had a half-life of just under a million years. Since the rate of radioactivity (A) is proportional to the number of atoms (N) but inversely proportional to the half-life ($A = 0.693 \times N/T_{1/2}$), this short half-life and large abundance meant that Urey could expect lots of heat production from ^{26}Al—but only if the planetesimals (i.e., the solar system) formed within a few million years of its creation.

Interesting idea, no? But difficult to prove, since ^{26}Al decays to ^{26}Mg, and magnesium is also a very abundant element, which means that the expected extra ^{26}Mg from aluminum decay would be lost amid the normal magnesium isotopes. Instead, in 1955 Gerry Wasserburg decided to search for evidence of ^{129}I decay in meteorites; if Urey was right and they formed soon enough for

^{26}Al to be present, surely ^{129}I would also have been there (since it has a longer half-life), and xenon is such a rare element that the expected radiogenic ^{129}Xe should stand out clearly.

But, as it turned out, no anomaly in the isotopes of xenon were found. The conclusion was that Urey was wrong: the time delay between nucleogenesis (the creation of the elements) and formation of the solid bodies of the solar system was too long for ^{129}I to have remained, and so much too long for ^{26}Al.

Normally one contradictory datum is not enough to throw away a lovely theory. (Remember, Einstein's theory of relativity was "proven" wrong by the first experimental test, only to flourish when others showed him to be right and the first experiment was shown to be flawed.) But most rare gas experimentalists were happy to be convinced by this one datum, because xenon is such a bitch to measure. It adsorbs onto the glass apparatus and hides during the run, then comes out at awkward times if a bit of heat touches the glass, and of course it's very rare, and the lower the number of atoms in a sample the harder they are to find and measure accurately. And very accurate data would be necessary to find the expected anomaly. So everyone forgot about it and got on with their lives.

Everyone but Urey, who came up with another suggested experiment: in iron meteorites, palladium should be greatly enriched over silver, and palladium had a short-lived radioisotope, ^{107}Pd, with a half-life of seven million years. No one was interested, and so two years later he carried out the experiment himself—and found nothing. No anomaly.

Well of course not, all the theorists agreed, happy with the data. There was no way to remove the elements from the stars where they were created, gather them together into a cloud of sufficient mass to condense itself into a solar system, carry out the condensation into various minerals and then somehow aggregate them into sun, planets, and meteorites without melting those primordial minerals, all within a few million years. No, Urey was wrong. There would be no isotopic anomalies in the solar system due to short-lived radioactivities.

At least that was the situation in 1959, when I was a postdoc at Brookhaven, just learning how to manhandle the mass spectrometer, and John Reynolds, then an assistant professor at Berkeley, came to visit. Our mass spec was based on his original design and was known as a Reynolds-type machine, so I was glad to meet him, but incredulous when he told us he had just found a ^{129}Xe anomaly in a meteorite. Frankly, I didn't believe him.

Since all the theorists agreed that there could be no short-lived anomalies, I had decided to try to find one. Having the rare gas machine, I looked for xenon, and having recently found traces of primordial helium in an iron meteorite—when none had ever been found in stone meteorites—I was certain that if the anomaly were to be found, the irons were the place to look. I looked, and hadn't found anything, so when Reynolds said that he had found it in a stone meteorite, I knew he must be mistaken. I didn't realize, at that early stage in my career, what a lousy experimentalist I was, compared to him, but I began to suspect something of the sort when he—mildly stung by my obvious disbelief—produced the proof. Figure 4 would soon be published in the *Physical Review*, showing very clearly a large excess at mass 129. I ran to the textbooks to try to find some contaminant with that same mass, perhaps an organic molecule that had strayed past the clean-up traps in his apparatus, but he was ahead of me, he had already done that, there was nothing at mass 129 but xenon, the fossil daughter of short-lived ^{129}I. Urey was probably wrong in the details of his model for solar system formation,[2] but he was right that it took place quite soon after nucleosynthesis—as all the theorists now ran around trying to prove, murmuring that they had known it all along.

2. In the past decade, we have found many planetary systems around other stars, and since they are so different from ours, and often from each other, we're at a loss to explain the details of how they—and we—formed. Urey's model, like every other one now being argued about, is probably wrong.

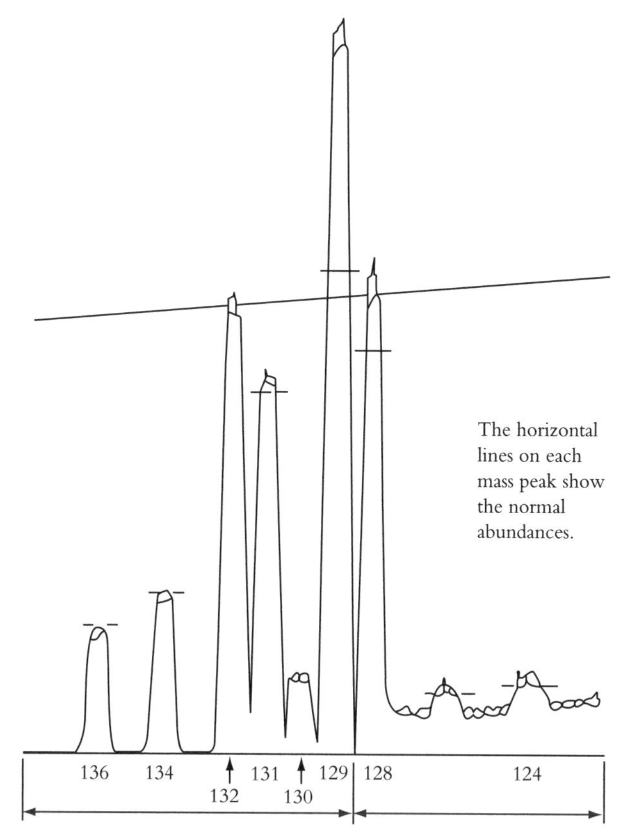

FIGURE 4. Mass Spectrum of Xe from the Richardson Chrondite

The first explanation offered was rather obvious, at least in retrospect. Iodine-129 is created in the tremendous flux of neutrons accompanying a supernova explosion, so clearly (read "perhaps") a supernova triggered the formation of our solar system, presumably by forcing the collapse of a preexisting cloud of gas and dust—into which the supernova debris, containing newly formed ^{129}I, was flung—and resulting in gravitational accumulation into sun, planets, meteorites, etc. Well, yes, perhaps. On the other hand...

It wasn't only the theorists who were scampering around. A host of experimental laboratories now plunged into the game of finding extinct radioactivities, and find them they did: isotopes of plutonium, palladium, samarium, silver, aluminum . . . the list goes on and on. (Just as once the "impossible" four-minute mile had been broken by Roger Bannister, everyone seemed to do it without difficulty. One of the mysteries of the universe, which poets seem to understand.[3]) As is usual in science, the multiplicity of data has raised more questions than it has solved, leading to theories involving "nucleosynthesis in a massive star dying close in space and time to the nascent solar system or production by local irradiation of part of the protosolar disk by high-energy solar cosmic rays"—or some other process as yet unknown. At any rate, more than one process seems to be necessary, as discussed in a recent paper; indeed, an entire conference was held recently to discuss the situation (see Notes). Complications, complications . . . what can you do? (Einstein was wrong: *der Herrgott* does seem to be malicious, at least to cosmochemists.)

Nevertheless, we surge forward. Not only do we now know when our elements were created, we're beginning to get a handle on the precise details of *how* they were created. "By applying theoretical constraints to three-dimensional fits of xenon isotope data from presolar grains," as a recent paper puts it, the details of the various stellar nucleosynthetic processes are beginning to be elucidated. We've moved far from the days of Lord Kelvin, when geologists were casually ignored by theoretical physicists.

Whatever the final result—hopefully, a detailed explanation of how both elements and planetary systems form—the time constraint presented by the xenon-129 anomaly was, to use Churchill's

3. "So easy it seemed/Once found, which yet unfound most would have thought/Impossible."— John Milton, *Paradise Lost*, Book 6.

expression, perhaps not the beginning of the end but at least the end of the beginning, in one step reducing to rubble all the previous notions and getting us started on a detailed, experimentally based investigation of nuclear astrophysics. It also started a nasty controversy in our old problem of earth history.

Claude Allègre, a French scientist, went to Berkeley to learn rare gas techniques from the master, John Reynolds. He then went back to Paris and put together a fine group of experimentalists concentrating on measurements of earth materials, particularly mid-ocean-ridge basalt (MORB). In 1982 they found evidence of a ^{129}Xe anomaly and promptly interpreted it in terms of the arguments we discussed in the previous chapter, applying it to the earlier versus later degassing scenarios: "Due to the short half life of extinct ^{129}I, this result clearly shows that the separation between the MORB mantle source and the atmosphere occurred 4,400 Myr ago; i.e., the atmosphere was outgassed and formed in the very early days of the Earth's history."

But it doesn't clearly show that at all. Imagine that the earth forms before all of the ^{129}I can decay—and incidentally, this is the one very strong point the Paris data makes. It ties the birth of the earth to the same time span as that of the meteorites and strengthens the supposition that the entire solar system formed at the same time, in contrast to earlier theories that supposed the solar system to have formed from the outside in (that is, the furthest planets formed first, and the rest in sequence). Okay, the xenon that is now in the atmosphere was initially trapped in the earth, and *some* of it comes bubbling out before the ^{129}I can decay (some "4,400 Myr ago"). The rest stays inside and the iodine decays to xenon, so when Allègre is finally born and learns mass spectrometry and melts a MORB sample containing this xenon it has a higher isotopic proportion of ^{129}Xe (the xenon anomaly) than the atmosphere. Voilà!

You might even be tempted to say *most* of the xenon comes out early, but the exact amount is model-dependent and the calculations are fraught with assumptions. I was able to show that the Paris data actually specifies only a range of 75% to 25% early degassing (with 25%–75% late degassing). As a later review stated, "Regardless of the exact degassing history used, this [the ^{129}Xe data] requires that strong degassing occur very early in Earth history," but the exact amount is still debatable.

And on the other hand, a 1999 paper in *Science* argued that you can't even say that much: "The variations observed in ^{129}Xe/^{130}Xe between . . . Earth's atmosphere and mantle samples may be generated by variations of iodine/xenon in terrestrial reservoirs, as opposed to rapid early degassing." Which is more in line with what the review quoted in the previous chapter stated, that other data "render invalid all calculations of degassing of the solid Earth."

Ah well. It's been a lot of fun, and that's not to be sneezed at.

Another bit of serendipity at the Reynolds lab has opened up a new and continuing line of research. The excitement generated by the ^{129}Xe excess would bring a long line of young researchers to Berkeley, each of them eager to learn meteoritic mass spectrometry. Among the first were Craig Merrihue and Grenville Turner, and they enthusiastically set to work determining the I-Xe ages of meteorites.

The mere existence of a Xe excess indicated that the solid minerals in which it was embedded had formed before the ^{129}I had decayed, that is, within a few iodine half-lives, or roughly less than a hundred million years. In order to quantify this time interval—to determine the I-Xe age—it would be necessary to measure the amount of ^{129}I in each mineral at the time of its formation, and that could not be done, because by now it was all gone, converted into xenon. But immediately following the discovery of the excess, Reynolds discovered a way to do this (following an earlier suggestion).

The stable isotope of iodine is ^{127}I, and there were good theoretical grounds for calculating the 127/129 rate of production in stars. So Reynolds irradiated the meteorites with neutrons, inducing the $^{127}I(n, \gamma)^{128}I$ reaction. ^{128}I soon decayed to ^{128}Xe (its half-life is twenty-five minutes), which was then measured in the mass spec together with the ^{129}Xe. By subjecting the sample to successive stepped temperature runs, he was able to generate a series of data, and when he plotted the ^{129}Xe against the ^{128}Xe (normalized to a stable xenon isotope produced neither by the irradiation nor by iodine decay), he obtained a straight line, proving that the ^{129}Xe actually did come from ^{129}I. Moreover, the slope of the line is determined by (and therefore determines) the I-Xe age. Eagerly, Turner and Merrihue, together with a colleague, Bob Pepin, got to work. Under John Reynolds' leadership they effectively established the field of early solar system chronology, so that now, instead of saying that the solar system formed about 4.5 billion years ago, we can say, for example, that a particular meteorite formed 4.5266 billion years ago.

But there was more to come, as serendipity now combined with cleverness. (Both are necessary: it's good to be good, it's better to be lucky, it's best to be both.) The simplest way to do the experiment would have been to heat the meteorite, freeze the released xenon down on charcoal, pump out the other gases, then heat the charcoal and measure the xenon. But Reynolds advised them always to measure all the noble gases, because "you never know what you might discover." This increased the work load considerably: they had to separate the different gases by differential temperature adsorptions on charcoal and then run each one, increasing the work load by a factor of two or three. And then there were all these other data to analyze, in which they weren't interested, and which had to be analyzed by eyeball on recording charts (today that's all handled by the computer, but then it was a laborious process).

As it turned out, Reynolds knew what he was talking about, for one day in 1963, Turner recalls, while Reynolds was out of town, "Craig Merrihue walked into my office to show me a chart from a neutron-irradiated meteorite which contained a tiny peak

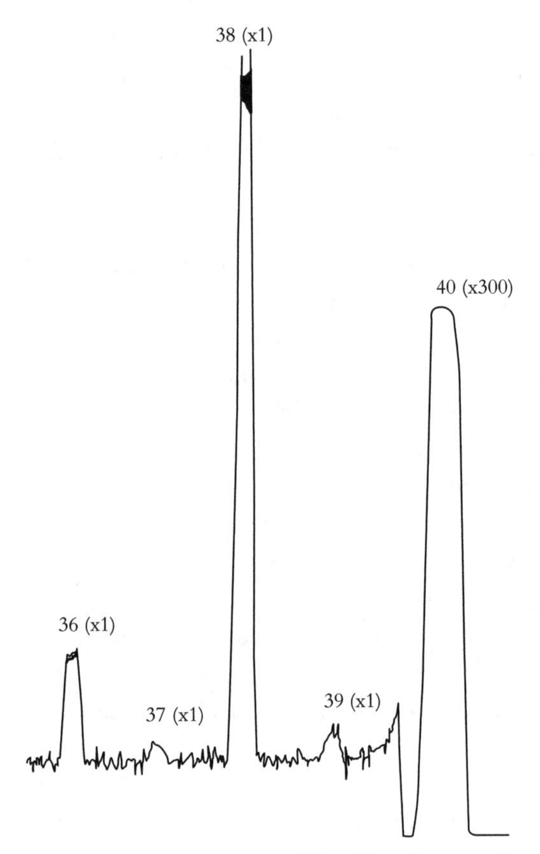

FIGURE 5. Mass spectrum of irradiated meteorite

at mass 39, between the usual argon peaks at masses 38 and 40 (figure 5). Merrihue had concluded that the peak was ^{39}Ar produced in the reactor by the reaction ^{39}K(n,p)^{39}Ar, and could be used to determine K-Ar ages."[4]

The thing to do would be to dredge through the argon charts from all their old I-Xe experiments and pull out the argon data,

4. This had actually been noted in two earlier reports, one in German and one in Icelandic. Neither of these were followed up with any real data, and had passed unnoticed by the scientific community.

but almost immediately Joe Zähringer's group in Heidelberg published K-Ar ages for a large number of meteorites (done the usual way, with an assumed average K abundance), and Turner and Merrihue decided that it just wasn't worth the effort to add a few more ages, since nothing unusual or of importance had come out of the Heidelberg data. (By this time everyone knew the meteorites had formed roughly four and a half billion years ago, and the K-Ar method couldn't be more accurate than that.) So Merrihue submitted an abstract outlining the new method of measuring K-Ar ages and went back to the new and exciting field of xenon.

The following year, Turner returned to England to set up his own mass spec laboratory in Sheffield. Since such a set-up procedure involves a lot of dead time, waiting for deliveries and repairs, Turner (instead of spending his time at the local pub) began corresponding with Merrihue about analyzing the old argon data in terms of potassium-argon ages. They were looking at the data for the Bruderheim meteorite, and they soon ran into a problem: the argon and potassium values sometimes correlated, as they should, but sometimes they just did not. As Merrihue wrote, "The enclosed argon correlation plots are baffling..."

That letter, written February 3, 1965, was his last. A month later Grenville received a letter from Bob Pepin telling him that Craig had died in a mountaineering accident. Sadly, still baffled, Turner put the data in a desk drawer and shut it...

Several months later he happened upon an article written by Ed Anders, who, upon reviewing a number of argon and helium ages, concluded that a high proportion of hypersthene chondrites—the class of meteorites to which Bruderheim belonged—had experienced heating and outgassing during a major parent-body collision 500 Ma ago.

It was an "Aha!" moment. Turner opened that desk drawer and "Looking at the Bruderheim graph in the light of this, I drew a line through the low temperature points... The slope corresponded to 550 Ma!"

And thus was born the field of 40/39 (or Ar/Ar) thermochronology. What Turner had discovered was a new way to date geologic events. K/Ar dating traditionally depends on an event in which the system to be dated suddenly lost all its argon; this is generally a melting event. But now Turner had discovered that a heating, nonmelting event would cause loss of argon from some minerals but not others. A stepwise heating experiment would reveal this, with the first (low-temperature) steps dating the partial-loss heating event. In practice, it has been possible to determine the thermal history of specific rock samples; for example, to determine the evolutionary history of the earth's crust by looking at 40/39 data on orogenic terranes.

Another development in the 1970s was the use of a laser beam to vaporize small areas of rock and thus to measure 40/39 ages in individual crystals, or even to control the laser beam to progressively heat an individual spot to elucidate its thermal history. Such studies have been particularly useful for the lunar rocks. These studies are covered in full detail in a book by McDougall and Harrison.

The importance of Turner's pioneering studies can't be overstated. When I took an informal survey of all noble gas laboratories last year, I found that more than half were involved in 40/39 work.

John Reynolds was[5] one of those scientists who showed that you could be fiercely determined and hugely successful in scientific pursuits and still remain a pleasant, unselfish, giving man. He was a young assistant professor of physics at the time of his discovery of the xenon anomaly, pursuing a line of research not shared by anyone else in the department, but with this discovery he transformed the field of meteoritics and attracted a series of excellent

5. He remained an active and dominant force in meteoritics until his death in 2000.

grad students and postdocs to his laboratory, ending up as a very popular head of the department and a leading researcher in the field he personally created.

Claude Allègre (on the other hand) showed that you could be fiercely determined and hugely successful in scientific pursuits *without* being a pleasant... etc. When I published the paper noted above, modifying and toning down his conclusions about early degassing based on his xenon data, he was furious. He wrote to the editor, excoriating him for publishing such an "absurd" piece of work: "I find properly scandalous that the paper by Fisher on rare gases was published... Such paper is so bad that I cannot make a comment myself!" [*sic*].

The editor replied with restraint, pointing out to him that three reviewers found the paper quite reasonable, and that a system was in place when scientists disagree: instead of vituperation, Allègre was free to write a "Comment" on the article, to which I would be free to "Reply." His colleagues wrote the Comment; it and my Reply were then sent out to reviewers and subsequently published. In all modesty, I think I demolished his/their arguments, though not their animosity. *Que sera, sera.*

Allègre went on to receive several major awards for his work; in 1997 he became Minister of Research and Education of France, but he was fired in 2000 when a coalition of French scientists and teachers protested his policies. "For many researchers and teachers, Allègre had become the man they loved to hate... with an aggressive, combative style." He is an opponent of working to offset climate change, charging in the French news magazine *L'Express* that "the ecology of helpless protesting has become a very lucrative business for some people."

Well, it takes all kinds. Unfortunately.

SIXTEEN

The Coldest Place on Earth

Directions for obtaining the Knowledge of all Dark Things.

—Rhind Papyrus, 1700 B.C.

SUDDENLY, AT 7.30 P.M. ON JULY 10, 1908, THE COLDEST place on earth—the coldest place in the entire *history* of the earth—was inside a small glass tube in a messy laboratory in Groningen, the Netherlands, where the temperature was a cool 269 degrees below freezing. That's centigrade; it would be minus 452° Fahrenheit. Inside the tube were 60 cc of liquid helium, produced for the first time in history by a Dutch physicist today virtually and unfairly unknown to the general public, Heike Kamerlingh Onnes—unfairly unknown, for unlike the results discussed in the last two chapters on MORB geochemistry, this feat of engineering physics has had profound practical consequences.

The utilization of fire was the first giant leap for mankind, but its opposite, the search for cold, has been an ongoing human activity through recorded history. (Actually, there is no such thing as cold; there are only lesser amounts of heat. Absolute zero, minus 273° centigrade, is unattainable, as "explained" by a complex quantum theory argument, and there are no minus numbers on the Absolute, or Kelvin, scale.) But in practical terms, no one cared, the important thing was to get ice for food preservation through the hot summers, and until nearly a hundred years ago the only way to do that was to bring it down from the high northern latitudes or, in the in-between latitudes, to store the winter's ice underground.

By the last quarter of the nineteenth century, some progress began to be made in utilizing that marvelous insight into nature, the Second Law of Thermodynamics, to bring some sort of mechanical cooling into people's lives. The law can be stated in various ways, but for this purpose the simplest is that heat flows from hot to cold. What could be simpler? And yet it has profound consequences.

When you want a cold drink, you put in ice cubes and the heat flows from the warm scotch to the cold ice cubes, cooling down the scotch. But as the ice melts, it dilutes the scotch, which is a problem. The solution? My wife bought a stainless steel cooling cube; the instructions said to put it in the freezer and use it instead of ice cubes.

But it didn't work. It didn't dilute the scotch, but it wasn't effective in cooling it down either. Why not? Look at figure 6, beginning at the left.

What we have here is the heat absorbed by water as it increases in temperature. At the left is solid water, ice, like the ice cubes in

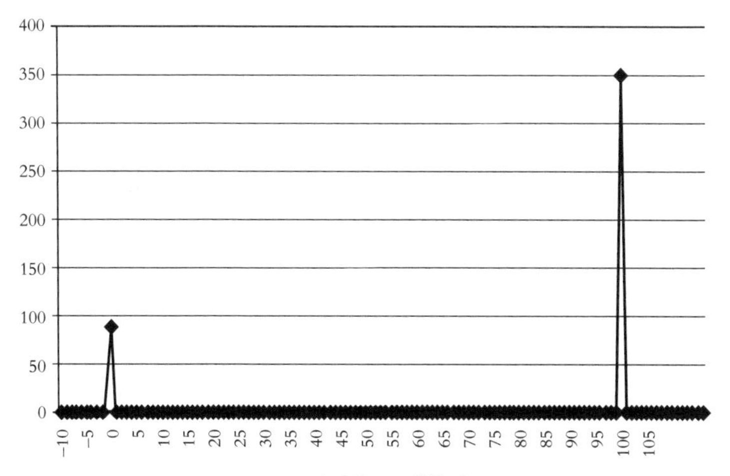

FIGURE 6. Heat of Fusion

the scotch, and one gram absorbs one calorie of heat from the scotch as it warms one degree of temperature. So going from minus ten to minus nine, one calorie of heat is absorbed. And so on and so on—until the ice has warmed to zero degrees. And now something funny happens: before it can warm to plus one degree it has to melt, and that takes a lot of heat. So going from ice at zero degrees to liquid water at zero degrees it absorbs 80 calories.

This is the heat of fusion, and it's why ice is effective in cooling down the scotch: when it melts it absorbs all that heat. The stainless steel works just as well in warming from minus ten to minus nine, etc, but the steel never melts. And so it just absorbs the heat one calorie at a time, nowhere nearly as effective as the ice. So you just have to live with your diluted scotch if you like it cold.[1]

It wasn't the idea of cold scotch (regarded by some as an obscenity) that inspired the inventors, but rather food preservation and, increasingly in the nineteenth century, the beer that was being produced by the lager brewers of continental Europe and North America. British beers were fermented (and drunk) at room temperatures, but the lagers were done at 5°–8°C (and their lack of taste means that they simply must be drunk cold, the colder the better, to numb the taste buds). In the summer, when demand was greatest, this meant a tremendous need for ice, and the growing demand for lager was one of the most persistent impetuses in the search for artificial refrigeration.

When would-be inventors looked at figure 6, they saw another, even bigger bump in the graph. When water reaches one hundred degrees, it boils, absorbing its heat of vaporization—more than 300 calories—and thus was born the idea of mechanical refrigeration.

The idea is simple. The first iceboxes were exactly that: you put a block of ice in an insulated box, and as the ice melted it absorbed heat from the air in the box, and so the inside air stayed cold. But

1. Okay, you could keep the bottle in the refrigerator, but that's missing the point I'm making here.

this meant lugging a fifty-pound block of ice up the stairs every few days. Now we wanted mechanical refrigeration, so if you want to cool off the air in a box, you put in a liquid that boils at a lower temperature than that of the ambient air. Heat flows from the air to the liquid, boiling it and cooling itself off. The (boiled) vapor is then pumped out of the box, taking with it the heat it absorbed. The heat is then released out of the box, and the vapor cools back down to a liquid, is pumped back into the box, and the cycle continues. The result is that heat is transported out of the box and released somewhere else; outside in air-conditioning, or into the kitchen in the case of refrigeration.

The pumping is taken care of by electricity; the big problem in the latter years of the nineteenth century was finding a convenient liquid, one that is cheap, not harmful, and vaporizes at the required temperature. If you wanted to cool a room from 200 degrees to 150, water—which boils at 100 degrees—would be perfect. But that's not the temperature range of interest. You generally want to cool a room, or a box, from ambient temperatures in the thirties (centigrade, remember) down to a few degrees or even to negative temperatures. The first electrical refrigerators used ammonia, sulfur dioxide, or methyl chloride, but each of them had problems. Ammonia and sulfur dioxide are both corrosive and thus likely to develop leaks in the apparatus, and they cause health problems when people are exposed to them; methyl chloride is less corrosive and thus less likely to leak out, but when it does it is not only toxic but flammable, even explosive.

In 1928, Thomas Midgley, an industrial organic chemist at the General Motors research laboratory in Detroit, discovered two gases with excellent refrigerant properties: dichlorodifluoromethane and trichloromonofluoromethane, which became famous as the first of the CFCs (Chlorinated and Fluorinated hydroCarbons). They were perfect, inert (i.e., noncorrosive), nontoxic, and nonflammable, and with them the refrigeration and air-conditioning businesses became everyday realities for hundreds of millions of people. Of course, it did turn out, over the next fifty years or so, that they had this little problem about depleting the ozone layer... At any rate they've now been replaced, the ozone layer is rebuilding, and everyone (especially DuPont) is happy.

But back in the earlier years of the twentieth century, Heike Kamerlingh Onnes was trying to move the whole business along, building, as science does, on previous work. Starting with Aristotle's description of the universe as being composed of earth, air, fire, and water, our understanding had gone on to break it up into energy (fire) and matter, with the original earth, air, and water now being called solid, gas, and liquid. The idea that every substance was one of the three was quickly challenged with the recognition that water could be all of them: ice, liquid water, or steam. As we progressed through the Middle Ages and the Renaissance, chemists succeeded in melting solids and vaporizing liquids simply by heating them; liquefying gases turned out to be harder.

But not impossible. The most obvious way was to cool them, but obvious does not mean easy. Lavoisier had written that if the earth were to shift its orbit out toward the edge of the solar system, not only would the extreme cold freeze our oceans, but the gases in our air would condense into "new liquids...of which we are now wholly ignorant." But without moving the earth and without electrical refrigeration, the coldest stuff around was natural ice, and that just wasn't cold enough. The next most obvious way was to compress the gases; by 1780 two French scientists had been able to increase the pressure in their apparatus enough to liquefy sulfur dioxide, and in 1787 a Dutchman had used the same technique to condense ammonia. Then nothing happened for another forty years, until Michael Faraday liquefied a variety of gases during two decades beginning in the mid-1820s. Such was his reputation for unfailing experimentation that the gases he failed to liquefy were then perceived to be "permanent" gases: oxygen, carbon monoxide, nitric oxide, methane, and nitrogen (the rare gases had not yet been discovered). It took another few decades for the "permanent" title to be broken. It happened in

1877, when a piece of apparatus broke in the provincial village of Châtillon-sur-Seine.

It's impossible to overestimate the importance of serendipity to the progress of science—serendipity combined with a sharp mind. Everyone knows how Fleming discovered penicillin when he looked in a dirty Petri dish, but it could have been discovered nearly twenty years earlier as people observed the bleeding statues of Europe.

It had long been reported that statues of the Madonna sometimes wept bloody tears, and that those of Jesus sometimes leaked bloody stigmata. These were worshipped as miracles, but the leaks were actually caused by a common bacterium (*Serratia marcescens,* for a time called *Bacillus prodigiosus* from the Latin for "miraculous"), which produces a red pigment. (Of course, it also happened in statues of generals on horses and naked ladies, but no one fell down on their knees in front of these.)

In 1911 a Polish student, Franciszek Groer, noticed that the growth of this bacterium was halted by the presence of mold; that is, penicillin. But his supervisor told him that this was just a natural phenomenon and didn't mean anything, which he accepted, and so another few thousand people died in the next seventeen years until Alexander Fleming noticed his moldy dish with the conspicuous (to him) lack of bacteria.

Back to 1877. In Châtillon-sur-Seine, a French mining engineer was trying to liquefy oxygen by the standard method of increasing pressure. Unfortunately, his glass tube couldn't take the pressure and cracked, releasing the oxygen. Most people, like Groer and his supervisor (and probably you and me), would have cursed, fixed the cracked tube, and started over. Fortunately, Louis–Paul Cailletet noticed *and paid attention to* a thin haze which suddenly appeared inside the tube as the gas escaped and as quickly disappeared again. Could it have been condensed oxygen?

Could be. In fact, it was, in the first soon-to-be practical example of adiabatic expansion of a gas: expansion with no transfer of heat between the gas and the rest of the world. Ordinarily, if you heat a gas it will expand. This is how you fill a hot-air balloon and make it rise. In other words, for a gas to expand it must use energy, which it gets from the application of outside heat. What if you can get the gas to expand without furnishing it with heat? It must use its own internal heat, and so it uses up that heat and cools down. Cailletet's oxygen was under high pressure, so when his tube cracked, the oxygen came shooting out, expanding so rapidly that there was no time for it to transfer energy (heat) from its surroundings. The result was a rapid cooling, taking it briefly below its condensation temperature. It condensed as dew on the glass for a minisecond, before soaking up enough heat from the outside to evaporate again. In that one brief minisecond, which most of us might have missed, Louis-Paul saw the future.

Or at least part of it. As so often happens, an identical result was being obtained by another scientist. Raoul-Pierre Pictet, in Geneva, used a closed circulation system to reduce the temperature nearly, but not quite, enough to liquefy oxygen. With this he condensed sulfurous acid and reduced its temperature to −65° by allowing some to evaporate, then condensed the evaporate under pressure and pumped it back into the bath, which was then used to cool carbonic acid, which was in turn allowed to evaporate, bringing its temperature down to −140°. In this bath, a glass tube with oxygen was cooled, then allowed to evaporate as per Cailletet, producing the same liquid oxygen haze.

This could all have been done some twenty-five years earlier when James Joule and Lord Kelvin (then William Thomson) discovered this process, now known as the Joule-Thomson effect, but, surprisingly, no one made practical use of it. But perhaps it's not so surprising. $E = mc^2$ was published in 1905, but twenty years later Lord Rutherford was still saying "Any idea of getting useful energy out of the atom is pure moonshine." And so it goes.

At any rate, with Cailletet's and Pictet's discovery a new line of research opened, and soon two Polish scientists were able to produce small vats of liquid oxygen, which, exposed in an open bath, boils at −183° centigrade, so it was now possible to do physics and chemistry experiments with equipment bathed in liquid oxygen, that is, at an ambient temperature of −183°.

Why, you might well ask, would anyone want to do that? Why not, I'm tempted to answer. Who knows what secrets lie waiting to be discovered in any new region of the universe?—and the ultra-cold was certainly a new region. But I must resist the temptation, for already there was a reason to investigate: it had been found that electrical resistance in a metal decreased as the temperatures were (modestly) lowered; in other words, the conductivity increased. This seems reasonable, for electric current is a flow of electric charge, and as the temperature is lowered, the normal shaking and bouncing motion of the metallic atoms would decrease, allowing the charge to pass through more easily.

But what if the temperature continues to drop? Eventually, at absolute zero[2] all atomic motion would stop and the electric charge could zip through with practically no resistance—except that at that temperature *all* motion must stop, which means no electric current at all. So as the temperature begins to drop the conductivity increases, as the experiments of the time showed, but clearly this must level off and begin to drop as much colder temperatures are reached (plummeting to zero at absolute zero.)

Clearly? Well, nothing is clear until experiment shows it to be so, but certainly the drop in conductivity was to be expected. The question in the minds of the scientists of the day was just when the drop would begin. And here Heike Onnes comes on the scene.

And once again serendipity rears its lovely head. Onnes was interested in quite another matter. Just one year before Cailletet's breakthrough, in 1876, a totally unrelated event of no great significance

2. This is before quantum theory told us that absolute zero is impossible to attain (and also before everyone was convinced that atoms existed, so the explanation is a bit anachronistic).

occurred: a young Dutch scientist, J. D. van der Waals, who had recently received his PhD, was appointed the first professor of physics at the University of Amsterdam, one of Europe's newest universities. His doctoral thesis, it turned out, was of tremendous significance, eventually earning him a Nobel Prize. It suggested, and with precise mathematics supported the notion, that all material substances are composed of molecules, themselves collections of atoms,[3] and that between the molecules a force exists which, while much weaker than the forces holding the atoms together within the molecule, are responsible for the physical state of the substance: solid, liquid, or gas.

Imagine a solid substance; ice is perhaps the easiest to visualize. The molecules of water are held tightly together by this intermolecular force (which we now call the van der Waals force). Now heat the ice; eventually, the shaking motion of the molecules, in response to the heat, is enough to break the van der Waals force and the molecules are free to move around—the solid has become a liquid. Heat it still further and the motion of the molecules becomes more frenzied; some of them acquire enough energy to break loose entirely and sail away, becoming a gas. In this theory anything can be a solid, liquid, or gas, depending on its temperature—

—And pressure. Imagine that you're heating the water to its boiling point, but in a tightly sealed pressure cooker. As the water begins to evaporate, more and more molecules become gaseous and so the pressure of the gas phase builds up, and you can visualize it pushing down on the water phase and not allowing any more molecules to break loose. So as you keep on cooking, the liquid water gets hotter; its boiling point has been raised because of the overlying pressure. Or if you're boiling water on a stove in Denver, where the air pressure is lower, the water boils at a lower temperature (which is why cooking recipes vary in such locations).

3. At this time, neither atoms nor molecules were universally accepted as real.

Anyhow, that's what van der Waals is predicting in 1876, and that's what his friend and colleague Heike Onnes is wanting to test. Some people do this by taking things to higher temperatures; he wants to go lower. And, unknown to him, that's where the fun begins.

But first the ship blew up.

Ancient history, as far as Heike Onnes was concerned; he never gave it a thought. But in 1807, nearly a hundred years previously, a merchant ship sailed into the port of Leiden and docked, its holds illegally crammed with gunpowder. At 4:15 P.M. on Monday, the twelfth of January, it exploded. It was the worst explosion ever in Holland, not to be equaled until the Luftwaffe made its appearance in the following century. Hundreds of houses were destroyed, the entire area surrounding the canal laid waste, more than 150 left dead and thousands more wounded.

But it all happened before Heike was born, and although his laboratory was built on the site of the disaster, it was not one of the things he thought about. He was not alone in this; by 1895 the area had been rebuilt, nearly everyone who had lived through the event was now dead and gone, no one remembered it.

It was remembered, however, in the ordinances of the city. It was spelled out there: no explosives were to be permitted within the city limits. Even had he known this, Heike would not have thought twice about it when the vandalism occurred, nor would he have hesitated to call the police. Late on a Sunday night, with the lab unguarded, a group of kids smashed a window, turned on the water and left it running, and in general raised a bit of a ruckus. Furious, Onnes called the authorities and demanded a police patrol; there were sensitive and dangerous chemicals in the laboratory, and who knew what might happen if ignorant youths started messing around in there?

"Dangerous chemicals?" the authorities asked. "Such as?"

Gases under pressure, for example, he responded indignantly. If mishandled they might explode—

Uh oh. Did Herr Doktor Onnes realize, the authorities asked, that across the street from his laboratory—in which, they were now informed, explosive material was stored—across the street was a nursery school with a hundred children? No, no, the situation was clearly intolerable. He must get rid of the explosives immediately.

He tried in vain to explain that he didn't have any "explosives," only gases which were perfectly safe in his hands. It was the vandalism that had to be corrected.

Right. Trying to argue with the authorities was as unrealistic then as it is now, and just as time-consuming. For the next three years, all his work was put on hold while he tried to explain what he meant while concurrently attempting to make the lab safer against intrusion.

Finally, after lengthy, ongoing, frustrating negotiations, the city fathers were satisfied and he was able to begin his work.[4] Quickly he made up for lost time, and once started, he progressed rapidly. Where previous investigators had worked singly, doing everything themselves, Kamerlingh instead set up a factory system. With uneducated but trained[5] and specialized young men (known as his "blue-collar boys") doing their jobs properly, a sort of mass production system was set up, and soon—using the adiabatic expansion of gases precooled by three cooling cycles—he was producing liquid oxygen. Not a thin haze of the stuff, but tons of it (well, anyhow, liters). The technique was to take a fairly condensable gas and liquefy it, then allow it to evaporate, thus cooling a second gas to liquid state. Then he'd allow the second gas to

4. The police/vandalism issue was resolved by never facing it, as is usually the case. ("What hump?" as Igor says in *Young Frankenstein*.) Today the Web is full of stories of vandalism and minor thefts involving such dangerous materials as radioactive sources, but nothing much ever comes of it.

5. He trained them himself, having set up a School of Instrument Makers for the purpose.

evaporate, cooling the third, and when this evaporated it brought the temperature down sufficiently to liquefy oxygen.

His aim in all this, remember, was to study the predictions of van der Waals's theory, and in order to do this he needed simple molecules so that the interatomic forces would be negligible (for he wanted to study the intermolecular forces). Molecules with only one or two atoms were what he wanted to liquefy, and these were the hardest to do: nitrogen, oxygen, hydrogen, helium, and argon all liquefy only at extremely low temperatures. So the attainment of ultracold was his Holy Grail, in which he was more successful than Percival. After oxygen, applying the same technique—using a more condensable gas to cool the next most condensable—he was able to liquefy nitrogen, and by 1906 he had even managed hydrogen, driving the temperature down to −253°C, just twenty degrees above the unattainable absolute zero.

And with this we finally get to the point: liquid helium.

Kamerlingh was not alone in this quest. For more than ten years Alexander Dewar in England and Karol Olszewski in Poland had been working on the liquefaction of helium—working and failing. Olszewski had concluded that it would be necessary to get below 2°K,[6] which he thought impossible. Dewar thought that with precooling using liquid hydrogen it might be possible to get down to 4° or 5°, but not lower.

And then, suddenly one day, the twenty-eighth of February, 1908, so easily that it seemed it could have been done eons ago, Heike turned a stopcock, allowing seven liters of helium gas to expand from 100 atmospheres pressure down to 1 atmosphere, and—he couldn't believe his eyes! Snowflakes filled the glass tube, swirling around. Snowflakes of pure helium, solid helium!

6. That is, 2° on the Absolute, or Kelvin, scale, where 0°K is −273° centigrade.

"Heh, heh, heh," smiled *der Herrgott*. "Not quite so fast, my friend."

So after announcing his find to one and all, and then carefully checking it (the reverse of the best process, but what are you going to do when all your dreams suddenly seem to appear?) he found—after six weeks of intensive work—it wasn't helium after all, but flakes of solid hydrogen formed from a tiny unsuspected hydrogen impurity in the helium.

Back to the old drawing board. His brother somehow found a few sacks of North Carolina monazite in Germany, and was sufficiently aware of his brother's interests to know that monazite is a uranium mineral containing helium. Kamerlingh set his group to work, heating the mineral grains until they released their helium—but releasing also lots of other gas. It took a team of chemists four months to clean it up, using basically the same techniques we still use today when measuring the rare gases in a mass spectrometer, taking advantage of the fact that the rare gases are the only gases that won't react with anything. So we, and they, cook the exhaled gases over copper oxide to oxidize them, cool them down to freeze out such stuff as carbon dioxide, then pass them over powdered charcoal to absorb whatever's left. Finally, a charcoal trap chilled with liquid oxygen absorbs all the other rare gases, leaving only helium.[7] And finally, on July 10, he produced the world's first liquid helium, taking it down to just below 4.5 °K, roughly −269 °C.

And now the fun begins.

With his factory system he was soon able to produce large quantities of liquid helium, and with this he was able to conduct experiments at temperatures never before reachable. Verifying the van

7. They also had to pray there was no neon in their samples, for neon would stay gaseous along with helium even over the cold charcoal. But there was no reason to suspect neon in the monazite, and as it turned out there was none.

der Waals theory at these extreme temperatures was fine, but the real fun came—as it often does—with the unexpected. Turning his attention to the problem of electrical conductivity, he found that it continued to increase as he brought the temperature of various metals down toward the liquid helium temperature. When would it reverse itself?

Lord Kelvin assured him that it would. In 1897 J. J. Thomson had proved that the flow of electric charge which was called electricity was actually a flow of particles called electrons, and Kelvin pointed out that the motion of any particle must stop at absolute zero. So at some point in the lowering of temperature, the increased conductivity must stop and reverse itself.

Slowly, carefully, Kamerlingh lowered the temperature, all the way down to that of liquid helium. Then, allowing the helium to evaporate and thus lowering the temperature still more—Wow!

At 4.2 °K the resistance of the metal abruptly ceased entirely, and the flow of electrons passed through undisturbed. The metal (mercury in this first experiment) "has passed into a new state, which on account of its extraordinary electrical properties may be called the supraconductive [sic] state." This time there was no mistake in the experiment, no unseen contamination; the only change to his pronouncement was eventually to call the new state "superconductive."

To visualize this effect, imagine rolling a ball across the floor. Newton tells us that a ball in motion will continue in motion indefinitely, but we know it won't, because (as Newton also told us) there is a force of friction which slows it down and stops it. In the superconductive state—totally unsuspected before its discovery—it's as if there was no friction to slow down the electrons, and so they just continue forever. An electric current flowing through a superconducting wire will last forever[8]—without any energy source. Like Ol' Man River, it just keeps flowing along.

8. There's no such thing as "forever," but current theoretical estimates are that it could outlast the universe.

The first round of excitement was purely theoretical, but no less exciting for that. The discovery of a completely new phenomenon (for which Onnes received the Nobel Prize[9]) stimulated researchers around the world, and soon one advance after another piled up, the most exciting of which was the discovery that certain materials showed superconductivity at temperatures higher than 4.2 °K. Year by year the temperature rose, and in 1962 the first commercial superconducting wire was produced by Westinghouse. In 1987 another Nobel was awarded for work raising the temperature to 35 °K, and soon after it rose to 92 °K, allowing the use of readily available liquid nitrogen to maintain it.

Imagine the possibilities!

Let's get rid of the first possibility before we start, because it's an impossibility. It's not a perpetual motion machine. You can't use this electricity to run your air-conditioning, for example, or an electric car, because then it's doing work and work uses energy, so it will run down just like any other electric power. But there are real possibilities just as exciting. Well, almost.

It's the connection between electric current and magnetism that is leading us, beginning with Michael Faraday, who, knowing from previous workers that electricity and magnetism were related, demonstrated how to create electricity with a magnet, although he didn't understand what either electricity or magnetism was. James Clerk Maxwell "explained"[10] it all with his electromagnetic equations, which showed, among other things, that an electric current generates a magnetic field about it. Now consider a current flowing in a superconductor. It will last forever, and so there will be a permanent magnetic field stationed around it. Put this current in an iron rail and put a magnet with the proper reversed polarity in a train, and the train could float above the track. The

9. Another Nobel was awarded in 2003 to two Russians, Alexei Abrikosov and Vitaly Ginzburg, for an explanatory theory.

10. The equations are perfect and beautiful and fit the data, but what exactly *is* an electromagnetic wave? At the end of his life Einstein said he still didn't understand it, nor (in all humility) do I.

trains of the future could go roaring along through thin air at undreamt-of speeds without the friction associated with wheels. (Or not. I remember a party conversation one evening at Cornell in the early 1960s, with Tommy Gold and Phil Morrison discussing what to do with the overabundance of energy we would have as more and more nuclear reactors come on line. There would be so much energy available that no one would know what to do with it, and we all had fun imagining the limitless possibilities. But then Pete Seeger began singing:

> I've lived all my life in this country
> I love every flower and tree
> I expect to live here 'til I'm ninety
> It's the nukes that must go and not me.

And the dream died under clouds of carbon dioxide and carcinogenic pollutants. So it goes . . .)

In a more practical vein, some of the biggest advances in medicine and basic physics have come with superconducting magnets, which make use of the free-flowing electrons to create incredibly strong magnetic fields. Not only are the magnets stronger than normal, but because of the free flow of electrons no energy is needed to keep them running, so their operation is economically viable, directly leading to two major advances.

In medicine, the superstrong magnets allow us to make use of the natural magnetism of atomic nuclei and the associated phenomenon of nuclear magnetic resonance to create Magnetic Resonance Imaging (MRI) of the human body. Atomic nuclei with an odd number of nucleons (neutrons and protons) have a finite magnetic moment, and so will interact with an applied magnetic field. The hydrogen nucleus (the proton) not only has the strongest effect but is also a basic component of most of the human body (in water and organic molecules), and thus hydrogen is the most-used element. When a magnetic field is applied, the protons can absorb or emit energy at a particular frequency, the "resonance frequency," which is proportional to the strength of the magnetic field. This energy is applied inside the huge magnet as brief pulses of radio waves. The protons absorb the energy, then release it at different rates, depending on the chemical microenvironment of the hydrogen atom. Thus, the radio signals emitted by protons in water, muscle, and fat, for example, are distinctly different. In MRI, the applied field is slightly nonuniform, so the resonance frequency depends on where the protons are located, leading to the concept of imaging. Given the difference in signals emitted from different types of bodily tissue, with the location of each signal identifiable because of slight differences in resonance frequency, a very precise image of any part of the body can be created with incredible anatomical clarity and precision.

Of course it's a bit more complicated than that, but for us the important fact is that the necessary resolution depends on just how powerful the magnetic field is, and superconductors, cooled by liquid helium, are what make it possible.

In physics, superconducting magnets are an indispensable part of the newest high-energy synchrotrons, which whirl particles around faster and faster until they reach energies high enough to break apart when they smash into each other. It's like a child breaking a watch to see what's inside. What they currently hope to find is the Higgs boson, a so far theoretical particle responsible for the biggest unknown in the universe: the origin of mass.

For if the universe began as a Big Bang, an explosion of pure energy, where did all the mass come from? Einstein has told us that mass and energy are equivalent forms of the same stuff, but how and why did some of the energy transform itself into mass? The answer, theoretically speaking, is the Higgs boson, the only elementary particle with a sound theoretical basis which has not yet been observed. The European Organization for Nuclear Research (CERN) has built the Large Hadron Collider,[11] a subterranean synchrotron seventeen miles in diameter, with the intention of

11. Hadron is the generic term for the subatomic particles made up of quarks and governed by the strong nuclear force. The proton is the basic hadron.

accelerating opposing proton beams to energies as high as 7TeV[12] (seven million MeV) and then crashing them into each other to see if any Higgs bosons pop out. This incredible energy is necessary because the Higgs has an anticipated mass so large that it will take more than 0.1TeV to form it, and so far experiments in that energy range have failed to find it. The energy available at the Large Hadron Collider means that if the experiments fail to find it, it doesn't exist, so the entire structure of nuclear particle physics rests on this experiment.

And the experiment rests on superconducting magnets, because as the protons whirl around increasingly faster, the magnets which cause them to bend around and stay in the LHC beam must become stronger and stronger, and only superconducting magnets can reach the required strength. And superconducting magnets need liquid helium, which brings us full circle.

As if to illustrate the importance of the liquid helium, just one week and two days after the LHC successfully ran its first proton beams, on September 19, 2008, a faulty electrical connection caused a "quench" in two magnet sectors—the magnets suddenly shifted from superconducting to normal, and instantly the normal resistance resulted in a sudden heating of the coils, which boiled off six tons of liquid helium before it could be shut down.[13] It took over a year to get the machine running again, at a cost of several tens of millions of dollars.

The phenomenon of superconductivity is actually just one aspect of a more general phenomenon: superfluidity, a state with essentially zero viscosity discovered first in 1938 when liquid helium was cooled below 2.2 °K. And so now we travel from gigantic

12. This term (TeV) is often defined incorrectly on the internet. Beware.

13. The LHC magnets need nearly a hundred tons of liquid helium to operate, a far cry from the hazy mist Cailletet first saw.

supersynchrotrons and millions of dollars to small, inexpensive experiments that may explain the origin of the universe. It began in a big bang, that part is settled,[14] but what came next? Cosmic inflation, which supposes that the universe underwent a period of extremely rapid expansion shortly after the Big Bang, is a leading candidate, explaining much of what we observe today, except for one minor flaw: it leads to a universe which is empty. No matter, no energy, no nothing.

In this scenario, the big bang eruption leads to a universal froth, as it were, a conglomeration of expanding bubbles, one of which is our singular universe. (The other bubbles become other universes; the whole is the "multiverse.") But the bubbles turn out (theoretically, of course) to be vacuums: empty. When they "pop," when the rapid acceleration ends and our current milder expansion begins, there's nothing there. So where did all the matter and energy of our universe come from?

A series of experiments has found that liquid helium-3, in its superfluid state, does an astonishing thing. Normal atoms in a normal fluid have their momenta aligned in random directions. But helium-3, passing into its superfluid state, spontaneously lines the momenta of all its atoms into one direction, in what is known as "symmetry breaking." The result is that out of chaos—a random distribution of momenta—a kind of order is produced. It's now thought that a similar phenomenon at the end of the inflationary period could have led to the creation (from "nothing") of the forces which now exist in our universe[15] (with the exception of gravity, which doesn't fit into any quantum description of the universe).

Finally, and strange as it may seem, the theory developed for the coldest substance—liquid helium—now helps explain the behavior of some of the hottest objects in the universe: neutron stars.

14. Yes it is.
15. And from there the Higgs boson creates matter.

Although helium doesn't become a superfluid until the absurdly low temperature of 2.2°K, the theory that was proposed in 1956 to explain such behavior also predicts that exotic matter such as the composition of nuclei and neutron stars could exhibit superfluidity at enormous temperatures of greater than a billion degrees.

Supposedly composed of free neutrons and neutron-rich nuclei, neutron stars randomly undergo mysterious "glitches" in which they suddenly spin faster and then slow down again. The glitches were at first explained as "starquakes," similar to earthquakes, with the surprisingly solid surface of the star rupturing, but this has been found to be untenable because insufficient energy is released. But a transition from one metastable energy state to another in the core of the star, allowed because of the superfluid characteristics of the core, can provide the necessary energy.

And speaking of stars...

SEVENTEEN

Back to the stars

Twinkle, twinkle, little star
How I wonder what you are . . .

FRITZ HOUTERMANS, A SWISS/GERMAN/AUSTRIAN/ Dutch/Jewish[1] physicist, was the first person to realize what makes the stars shine.

Well, to tell the truth, "the first person to . . ." is a phrase badly used in science; it's often not exactly the truth. All scientific progress builds on a growing body of knowledge, and when that body grows to a certain level it sets up the next discovery for whoever is bright enough to grasp it. And frequently that means more than one person, so coincidental discoveries by more than one person are often the case.

Thus Lise Meitner discovered nuclear fission, which was hiding in the radiochemical data of groups working in Rome under Enrico Fermi and in Berlin under Otto Hahn, but so did another female German scientist, Ida Noddack; Meitner gets the credit,[2] Noddack is forgotten. The list of such simultaneous discoveries

1. Born in Germany of Dutch ancestry, one-quarter Jewish, brought up in Vienna, ended in Bern.

2. She gets the credit, but not the Nobel Prize. The Physics committee said her work was properly chemistry, and the Chemistry committee said it was undoubtedly physics, so neither nominated her, though everyone agreed that she should get the prize.

goes back a long way, through special relativity (Einstein and Lorentz) and calculus (Newton and Leibniz) and gravity (Newton and Hooke) all the way back to whoever were the first people to realize the earth was round.

So Fritz Houtermans actually may not have been the first, but no one preceded him, as far as we know. In 1959, at Brookhaven, while we were discussing a possible research appointment at his Physikalisches Institut in Berne (which he said he was laboriously tugging into the twentieth century), he told me of his epiphany. He was courting a lovely girl, he said, and in lieu of a local movie theatre he took her on a long walk into the countryside. Night fell and the moon came out and they lay down on a small rug he had thoughtfully (and hopefully) brought along and by the light of the moon they made love.

Afterwards he dozed, and woke to find the moon had set and the cool clear night was ablaze with stars. He lay there staring up at them, oblivious of the young lady curled beside him, and (he said) with an inspiration so sudden it was almost as if the stars themselves were telling him their secret, he realized what it was: "Helium, that was the key!"

He must have laughed aloud with joy, for the young lady rolled over on him, stroked his face lovingly, and asked what he was thinking—and, idiot that he was (as he said), he told her: "I know what makes the stars shine!"

She sat up angrily; *that's* what he was thinking about? She pulled her clothing together and set off down the mountain, with poor Fritz following behind, trying to explain.[3]

Poor Fritz indeed. Aside from losing the affections of that woman, he is noted as one of the unlucky people to be arrested and tortured by both the NKVD and the Nazis.

3. He evidently told the story slightly differently at different times. For another version, in which he later marries the young lady, see Khriplovich.

Following his upbringing in Vienna, he returned to Germany for his PhD in physics. Being both partially Jewish and an outspoken Communist, when Hitler came to power in 1933 he fled to England, where his Viennese tastes came into conflict with English cooking: "boiled mutton and salted potatoes" induced him to leave within the year, this time to Soviet Russia. He worked at the Kharkov Physico-Technical Institute for three years, but his German background suddenly became important when the Stalinist purges began, and he was fired and forbidden to find other work. He was granted an exit visa, but when he tried to pack up he was forbidden to take his books and notebooks. Not immediately realizing the seriousness of the situation, he spent the next two weeks arguing with the authorities; when he finally decided to leave his books and go, it was too late. He was arrested, imprisoned, and violently tortured to extract a confession of being a German spy. He resisted the physical torture, but when he was told that his wife would be arrested and his two children taken and sent with new identities into the void, to be forever lost to him, he falsely confessed.

Surprisingly, the result was that he was then released for deportation to any country of his choice. He suggested England or Sweden, but instead was sent to Germany—where he was immediately arrested on suspicion of being a Soviet spy! He lived through the Gestapo interrogations, managed finally to convince them of his innocence and ended up working on the German atomic bomb—and sending warnings through Switzerland to his friends in America: "Hurry up! We are on the track."

But the subject of the German atomic bomb is beyond the scope of this book. Instead, let's go back to 1928, when Fritz, finishing his PhD in Göttingen, met the Russian émigré George Gamow and encountered the helium-decay mystery.

Rutherford, you will remember, had shown that the so-called alpha particles emitted in the radioactive decay of heavy elements were actually helium nuclei. The mystery was, how did this happen? One could calculate from basic classical physics that there existed a Coulomb barrier forbidding the alpha particles (the helium nuclei) from getting out of these heavy nuclei.

The classical description sees the alpha particles bouncing around in a potential well inside the nucleus and coming up against the Coulomb force that exists between the positively charged protons of the nucleus and those of the alpha particle (which is nothing more than a doubly charged helium nucleus). This repulsive force can be calculated, and it turns out to be greater than the energy of the emitted alpha particles. In the isotope polonium-212, for example, the potential barrier to alpha emission can be calculated as 26 MeV, while the alphas actually come out with energies of only 8.8 MeV. So how can a "helium particle" (as Rutherford called them) of energy less than the barrier penetrate through this repulsive force and reach the outside?

The situation is analogous to attempting to throw a ball over a wall. If thrown hard enough (given enough energy), it can be thrown over the wall, but if it doesn't have at least that amount of energy it will hit the wall and bounce back. The alpha particles bounce against the Coulomb wall with insufficient energy to pass over it, and yet somehow they get through. It's as if the thrown ball hits the wall but instead of bouncing back it just slips through! It's impossible—but it happens.

Thus, in 1928, a mystery.

Which Gamow brilliantly solved, applying the newly discovered Schrödinger wave equation. This describes particles such as the alphas with a wave function; the result of the equation gives the probability of finding the particle in any given state. When setting up the equation for a particle of energy E coming up against a potential wall, where E is less than the energy of the wall, the astounding result is that the probability of the particle (wave function) bouncing back is large, but not quite

100%: there is a small but finite probability that the wave function exists—and therefore the particle exists—on the other side of the wall. It has somehow passed through the impenetrable wall. The result is intuitively ridiculous, but Gamow showed that the precise calculations matched the experimental results perfectly, and so the concept of "quantum tunneling" through a barrier was established.

Gamow had just arrived in Göttingen, coming from Leningrad, and he and Houtermans immediately "clicked... They were very much alike, both a bit Bohemian and reckless..." and in conversations at the local cafes, over tables littered with coffee cups and the Göttingen version of Sachertorte, Houtermans pounced on the tunneling concept and quickly realized that it allowed for the inverse: alpha particles from outside the nucleus could tunnel into it, and thus was born the concept of induced nuclear reactions.

More an experimentalist than a theoretician, he tried right away to calculate what energy would be necessary to carry out the experiment his idea suggested: throw an alpha particle at a nucleus and see if it actually tunnels in to form a different nucleus. Sadly, the result was that the necessary energies were beyond the capabilities of any equipment in existence at that time. But then came that happy star-filled night in the countryside outside Göttingen when, looking up at the stars, he realized that the heat in their interior would provide the energy needed to induce his hypothesized nuclear reactions, and—just as a single match can release the energy to start a fire that can consume a whole building—the nuclear reactions would release enough energy to keep the stars shining for the billions of years needed to satisfy Rutherford's determination of the age of the earth.

It was indeed an exciting moment. And he was right, and he was wrong. Nuclear reactions induced by the original heat of gravity in the core of stars do provide their energy, but not in Houterman's original sense; at least, not in the majority of stars.

These, generally termed "main sequence" stars, which constitute roughly 90 percent of those observed, get their energy not from helium-induced reactions but from those which *produce* helium: by a variable sequence of reactions, four hydrogen nuclei fuse together to form one helium nucleus. The resultant helium has less mass than the four hydrogens by about 0.7 percent, and this missing mass, converted into energy according to $E = mc^2$, is sufficient to provide the star's output of energy for billions of years. But it was another decade before this was understood, primarily by Hans Bethe.

And Bethe too was wrong as well as right. He had the idea right, but the details of the process he suggested were wrong. He thought the fusion of hydrogen into helium had to be catalyzed by carbon and nitrogen, in what is called the C-N cycle; it was later shown that direct fusion, or "burning," is the main process.

That might be a minor detail, but Bethe was wrong in a more serious sense. His calculations showed that the nuclear reactions stopped with the formation of helium, thereby shutting the door on what would later become known as stellar nucleosynthesis, the creation of all the elements of the universe by nuclear reactions in stars.

Gamow had suggested earlier that the roughly one hundred different elements were all formed during the creation of the universe, in what is known today as the Big Bang, by the continuous fusion of hydrogen, building up all the elements one by one. And indeed such fusion can produce elements of mass two, three, and four—but it ends abruptly at mass four with ^4He. Because there simply is no element of mass five. A hydrogen nucleus (a proton) trying to fuse with ^4He just can't do it, and Bethe argued that this same barrier exists in stars. And for the majority of stars, the main sequence stars, he was absolutely right: no other nuclide formation takes place.

But there are other kinds of stars, in particular the red giants. These have much hotter cores, and so a different nuclear reaction is necessary to account for them. Perhaps helium burning?

And so it was suggested that perhaps two helium nuclei could fuse, producing an element of mass 8, jumping the mass 5 barrier, except that there is also a blank spot at mass 8—no such nucleus exists.[4]

The breakthrough came from a younger colleague of Bethe's at Cornell, Edwin Salpeter. I met him at one of Tommy Gold's parties in Ithaca in 1963. He was offered an endowed professorship at Princeton but turned it down because Princeton's nepotism rules would not allow them to offer his wife Miriam, known as Mika, a job. Mika was a brilliant biologist, specializing in scanning electron microscopy, who had been relegated at Cornell to a continuing postdoctoral position instead of a professorship because of Cornell's own nepotism rules (which were somewhat more liberal than Princeton's).

The '60s were a time of change, but slow change. The physics building at Cornell, for example, had separate lavatories for male and female students, but only one for faculty: it had never occurred to anyone that there might be a female professor.[5] A few years after we left in 1966, Cornell brought in an outside board of review to look over their biology department, and their recommendation (I was told) was to fire the entire department—except for Mika Salpeter, who should be made a professor immediately. To their credit, Cornell followed the recommendations enough to put her on the professorial track. She died in 2000, shortly after their fiftieth wedding anniversary; today the Society for Neuroscience and Women in Science annually award the Mika Salpeter award for lifetime achievement.

4. Helium-8 has recently been created, but there seems to be no way to produce it in stars.

5. When I came to the Marine Lab at Miami in 1966, they would not admit female graduate students. When I asked why not, they patiently explained that there were no lavatory facilities for women on the research ships. It took another few years before someone had the bright idea to print a "Women" sign to be hung on the lav when a woman wanted to use it. Today more than half the grad students there are women. Sic transit et cetera...

Okay, back to the helium problem in stars. The idea was that all the elements heavier than helium could not have been produced in the Big Bang, as Gamow had hoped, because of the "missing mass barriers" at five and eight. And yet the heavier elements existed. Where else in the universe could serve as the furnace to cook them up? Where else but in stars, the hottest, most energetic spots in the universe? But how to get past the missing-mass barriers?

Why, with helium, of course. Main sequence stars eventually grow into red giants, pouring out so much energy that it was clear they needed another reaction to fuel them; hydrogen burning just wasn't energetic enough. Their interior is hot enough to fuse helium into beryllium, but the problem was that beryllium-8 is unstable and would break apart into the original two heliums; thus the missing-mass barrier at mass 8. But in 1951 Ed Salpeter visited Caltech to talk about nuclear astrophysics and found that an experimental group headed by William Fowler had just measured the beryllium-8 instability precisely, and Salpeter immediately pointed out that if they were right the mass-8 barrier wasn't quite complete: using the Fowler data, if two heliums fused to form beryllium-8, its decay would take roughly a 10^{-16}th of a second, an incredibly short time but one that might be long enough for another helium to come along and convert the not-yet-decayed beryllium-8 into the stable-nucleus carbon-12. In Fowler's words: "Salpeter immediately realized that that amount of instability was small enough that in red giant stars there would be a high enough concentration of beryllium-8 constantly being made, constantly disintegrating, that it could be hit while it was beryllium-8 by another helium nucleus to make carbon-12."

Alas, it seemed it was not to be. The production of ^{12}C thus calculated was too small by many orders of magnitude to fit the astronomical data—until the British astrophysicist Fred Hoyle agreed with Ed that it had to be, since there just wasn't any other way to make all the elements, and that it could be if the

$^{8}Be + {}^{4}He \rightarrow {}^{12}C$ reaction could be speeded up, which would happen if that ^{12}C nucleus had a resonant state at just the right energy. It's a complex quantum argument, but it's equivalent to opening another lane on a crowded highway: the cars immediately begin moving along faster.

The energy of the needed carbon state was calculated, and experiments by Fowler's group in California soon found that it indeed existed. So the Salpeter-Hoyle idea carried us past the missing masses, and from there on it's pretty straightforward. The simplest way would be just to add a hydrogen nucleus (protons) to change the carbon into nitrogen, then another to change the nitrogen into oxygen, and so on, but the Coulomb barrier prevents this. Instead, the heavier elements are created one at a time rather routinely in red giants and supernovas by the equivalent process of addition of neutrons followed by beta decay.

But, aha, it's not as simple as it sounds. To get some experimental data to work out the details of stellar nucleosynthesis, as it's called, we start with a sigh of disgust.

The sigh came from a scientist at the University of Liverpool who was passing through Miami one day in the latter 1960s. He was returning to England from a trip to the Barbados, where he had set up a station to collect space dust.

The idea at the time was that interplanetary space must have dust grains whirling around, some the result of asteroidal and meteoritic collisions with each other and with the terrestrial planets, some comet debris, and perhaps some of the original stuff left over from the formation of the solar system.[6] Just as meteorites

6. This would be the stuff from which the planets eventually coalesced. If it could be found, it would provide invaluable clues regarding solar system formation.

fall on the earth, so would this dust. In fact, the so-called shooting stars are precisely this: specks of dust falling through the atmosphere and burning up from the friction as they do. So David Parkin had set up a collecting station to find any of the dust grains that made it through intact. The problem was that normal earthly dust is ubiquitous, so he had located his station far away from any center of industry, off the shores of Florida, where no factories disturbed the atmosphere and where the prevailing winds came from the vast expanse of the Atlantic Ocean.

Unfortunately, the problem was greater than the solution, ergo the sigh of disgust. Parkin had found, upon inspecting the dust he had collected, that it was overwhelmingly nothing but terrestrial junk; nothing cosmic could be discerned. What he had discovered was that dust from the continents is blown all over the world, rendering his idea useless. Over a couple of beers at the Marine Lab, he offered his station to anyone who wanted it.

Once again, serendipity reared its head. A couple of years previously, Miami had recruited a bright young radiochemist who was just getting his PhD from Princeton. It had been thought that, aside from those we were producing in our cyclotrons and reactors, the only radioactive atoms on earth were those of potassium, thorium, and uranium and their daughters. These were left over from the creation of the elements, which took place before solar system formation, because the K, Th, and U half-lives were billions of years; any radioactivities with shorter half-lives would have died out by now. But in 1949 Willard Libby discovered that carbon-14, whose half-life was only thousands of years, was being produced constantly by the action of cosmic rays on the atmosphere. Fritz Koczy, the head of physical sciences at the Marine Lab, thought that aluminum-26 also would be formed by the action of cosmic rays on terrestrial materials, and would collect in oceanic sediments. Knowing its half-life (1.05 million years), it might be possible to measure the sedimentation rates by looking at this cosmogenic aluminum-26 as a function of depth. (The carbon-14 half-life limits work to a few tens of thousands of years; for most geological work, the time scale needed is millions of years.) Joe Prospero, the Princeton grad, thought it sounded good, and so he came to Miami and set up the equipment and looked at some of the sediments the Marine Lab's expeditions had collected, and found nothing.

The rate of production of the nuclide and its concentration in the sediments were too low for the apparatuses of the day. So Prospero was looking for something else to do, and Parkin's disgusted discovery of the ubiquity of terrestrial dust sounded as if it might be interesting. And so it has proved. Prospero took over the station, added more of them around the world, and parlayed the study of terrestrial dust into a prosperous (sorry about that) career spanning four decades and still going strong.

Meanwhile, Professor Don Brownlee of the University of Washington was looking for cosmic dust in a different way. He used balloons, rockets, spacecraft, and high-altitude airplanes to collect it before it drifted down low enough to be contaminated with earthly junk. In 1970, using a balloon-borne collector known as the Vacuum Monster, he was finally able to firmly identify cosmic dust as such. Since then, he has found the stuff on earth in polar ice and deep-sea sediments. This cosmic dust, which comes mainly from comets, has been invaluable in interpreting their origin and importance, and led more than a hundred scientists to cooperate on a mission to go straight to the source.

On the fourth of January, 2004, an unmanned spacecraft named Stardust rendezvoused with comet Wild 2 and collected more than 10,000 dust particles from its coma, returning them safely to Earth two years and ten days later. Noble gas measurements now show that these are primitive, unaltered grains that formed in the solar nebula, close to the sun while it was still evolving into its present form, and as such, they present a series of original photographs, still to be developed, recording early solar system evolution.

As important as this is, an even more exciting possibility for extraterrestrial dust was in the offing, not envisaged by the human imagination; it needed a little serendipity of its own, for which we return to John Reynolds, meteorites, and the xenon isotopes.

As far as we knew, although elemental abundances varied throughout the solar system, the isotopic ratios of any element were identical except for the effects of cosmic rays or radioactive decay; in fact, this was "one of the few assumptions that can be considered well-justified and firmly established," according to an early review. As late as 1996, a review stated, "Solar system materials—even primitive meteorites, which contain the oldest solar system objects—have uniform isotopic compositions." The model in everyone's head was that all the objects in the solar system formed from a "well-mixed primordial nebula of chemically and isotopically uniform composition."

But after Reynolds' identification of the xenon-129 anomaly, he and other workers following his lead in the 1960s and for the next two decades found other isotopic discrepancies in xenon (and later in neon), to the puzzlement of all.[7] The first anomaly was found by Reynolds and Grenville Turner when they heated a meteorite in several steps and found that one temperature fraction was enriched in the heavy isotopes. The effects were small, but repeated measurements left no doubt they were real, and explanations proliferated. When spontaneous fission of uranium failed to match the data, other, more imaginative, suggestions followed: perhaps mass fractionation during their incorporation into the meteorite parent bodies and the earth, perhaps the result of fission of an extinct nuclide or an undiscovered superheavy element, perhaps nuclear reactions as a result of an early solar system bombardment with cosmic rays or neutrons, perhaps all or none of the above.

And time marched on, and another xenon component was found to be enriched in the light isotopes. Since both components

7. That is, to the puzzlement of all who paid attention. Almost no one aside from the noble gas investigators paid attention; most astronomers simply didn't read the same journals.

were found in the same meteorites—marvel of marvels!—they must be related. But what could enrich both light and heavy isotopes? It was time to rethink the whole problem.

Slowly, the rethinking focused on a truly remarkable possibility. The whole concept of isotopic constancy within the solar system rested on the supposition that the elements were formed in stars, collected in a cloud or nebula, and stirred around thoroughly for millions of years before coalescing into the solar system. But what if some strain from another star or stars were slipped into the mix?

Since the noble gases are gases, any such strange stuff would have been well mixed in before the solar system formed. But what if some tiny grains of solid minerals from another star were tossed in, within which some xenon was trapped? This xenon might well have undergone a different mix of nuclear processes than the average stuff of our solar system, and might thus have a different isotopic composition. A few astronomers began to be titillated by the possibility, and gave it some serious thought. Perhaps a late supernova erupting just before solar system formation? Perhaps . . . ?

Never mind. Theories are fun, but data are facts. In Chicago Ed Anders tried a different tack. While other labs were trying to separate various xenon components by temperature, he organized a series of experiments to locate the mineral that was carrying the xenon. Taking inspiration from the Curies, who dissolved tons of pitchblende to extract a few grams of radium, he began dissolving the meteorites piece by piece. The technique has been compared to the best way of finding a needle in a haystack: instead of looking for the needle, burn down the haystack, blow away the ashes, and there the needle is, left behind.

By 1975, successively dissolving various minerals and testing the residue for xenon, redissolving and retesting over and over again, he had reduced the meteorite to less than 1% of its original mass and ended with a conglomerate mineral that was responsible for the anomalous xenon. By 1987 he had identified the specific mineral holding the xenon: it was diamond,[8] and comprised less than 0.04% of the meteorite. In the next twenty years, several other mineral grains have been identified and other isotopic anomalies have been found, both in cosmic dust and in meteorites, the only explanation of which is that they formed in a variety of stars under different nucleosynthetic conditions; they have become known collectively as stardust. With them, for the first time we are able to study precisely what goes on in the different phases of a star's life, since the various processes that create the isotopes are the lifeblood of stars as they pass from the main sequence into their other phases: red giant, white dwarf, supernova, and so on, each with its own blend of isotopes, leaving a record to be read here on earth.[9]

Finally, to close the circle, remember that we began with Prospero's failed search in the 1960s for cosmic ray–produced aluminum-26 on earth with which to measure sedimentation rates. It took another couple of dozen years to find a suitable cosmogenic nuclide on earth, and then it wasn't aluminum but an isotope of krypton. Another decade passed before a sudden flood of cosmogenic noble gases were found, which today are being used, together with the mantle gases, to study not just sedimentation rates in the oceans but everything from the history of water in

8. A curious coincidence. Anders had long been interested in meteoritic diamonds, which were taken by some to be proof of meteorite origins in a planetary-sized body (thought to be necessary to provide the high pressure needed to form diamonds, as on earth). Anders showed instead that the (tiny) meteoritic diamonds were formed in the high pressures caused by shock events (collisions of small bodies). I once heard him bring a UCLA graduate student to tears in a filled conference hall with his unremitting data-supported cross-examination, which demolished the student's suggested planetary origin.

9. The record is a complex one, not yet completely disentangled; it will provide fun and funding for years to come, but is beyond the scope of this book. See, for example, J. D. Gilmour and G. Turner, "Constraints on Nucleosynthesis from Xenon Isotopes in Presolar Material," *Astrophysics Journal* 657 (2007): 600–608.

the Sahara to the accumulation of ice in the arctic and practically every geologic process occurring in between: the rise and fall of continents, the ebb and flow of lavas and glaciers, the accumulation of soil and the uplift of oceanic islands, and—by measuring the formation of deep oceanic waters and their global circulation—the rather important (in these days of climate change) interrelationship of the oceans and our climate.

And with new techniques and sophisticated instrumentation that weren't even dreamed of when Prospero ended his search, ^{26}Al too has been found. Ironically, it doesn't seem to have its origin in cosmic ray effects on terrestrial rocks, as Koczy and everyone else had assumed, but in cosmic ray reactions on atmospheric argon. And so we begin and end with the noble gases.

EIGHTEEN

The Neutrino Revolution

They all laughed at Wilbur and his brother
When they said that man could fly . . .

—I. Gershwin

But while all this was going on, while the noble gases were being used to work out all the details of stellar processes, a different argon-based experiment was sneaking in and threatening to upset the whole applecart. I first began to learn about it way back in the fading summer of 1958, when I pulled myself up off the Westhampton sands and sauntered back to the lab, angry—in my own self-importance—that Gert Friedlander had hopped off to Europe and left me on my own. You'll remember Ray Davis, in whose lab I was to work on the iron meteorite K/Ar problem? Well, I first met him that summer when I found Ollie Schaeffer and his mass spectrometer. In the lab next door was this courtly, soft-spoken Southern gentleman, Raymond Davis, Junior, who was putting together a most unlikely experiment and who invited me to join him in his journey into the unknown.

Except that it wasn't really unknown. It was a basic part of quantum mechanics, the theory describing the inner workings of atomic nuclei, which was put together largely during the 1920s and '30s—some thirty years before my sojourn at Brookhaven, and which I considered a time of ancient history, not quite real. Oh, I accepted that the 1920s had really existed, but in an intellectual way only, as a sort of existential fantasy—they had happened before I was born. (I first noticed this in others when, in the 1980s, I referred during a class lecture to the Kennedy assassination and was received with blank, uninterested stares. The students knew about it, but it had happened before they were born and had the same status as the Lincoln assassination: it was true, certainly, but basically it was a story grown-ups told.) It's hard to realize that I'm writing this now more than twice as far removed from my Brookhaven years as those years were from the beginnings of quantum mechanics.

So anyhow, it was known back then that the nuclei of atoms were held together by a binding energy which can be expressed through Einstein's famous equation $E = mc^2$. In this case, the mass is the difference between the actual measured mass of the nucleus and the masses of its constituent protons and neutrons. The nucleus has less mass than its component parts, and this "missing" mass is the energy in Einstein's equation, holding it together.

In the simplest form of the radioactive process known as beta decay, a neutron changes into a proton plus an electron which is emitted from the nucleus, carrying off energy. Consider a set of nuclei which have the same total number of neutrons and protons, but varying numbers of each. For example, the nuclei of carbon-14 and nitrogen-14: the former has six protons and eight neutrons, the latter has seven of each. The carbon has more mass (therefore more energy) than the nitrogen, enabling it to beta-decay into nitrogen, with the excess energy carried off by the electron.[1]

1. Since one of the carbon's neutrons becomes a proton, its nucleus now has seven protons and seven neutrons; in other words, it's now a nitrogen nucleus.

By the 1930s, the physicists knew how to measure precisely both the difference in binding energies of carbon and nitrogen and the energy carried away by the electron. And thereby arose a problem: the electrons in any radioactive decay came flying out not with a unique energy corresponding to the difference in binding energies, as they ought, but with a smeared-out spectrum of energies. True, the highest energy corresponded to the difference in binding energies, but most of the electrons were emitted with less energy than that.

In other words, some energy had simply, unequivocally, and absurdly disappeared. Where did it go? Energy cannot disappear! And yet it seemed that it did.

This was big stuff. Niels Bohr, in a sort of controlled panic, suggested that we discard conservation of energy, one of the firmest bases of physics. Energy couldn't be created, he argued, but it sure looked like it could disappear. A different but just as controversial answer was provided by Wolfgang Pauli, who suggested that in the process of beta decay a new particle was formed. Since it had never been observed, he concluded that it had no electrical charge and little or no mass, so that as it flew away from the decay, carrying the missing energy, it slipped through any measuring equipment without leaving a trace.

Pauli was an otherworldly Austrian physicist. A pudgy little man, he continually rocked back and forth, even while walking, which made for an unusual gait: as he rocked forward he'd take quick little steps to catch up with his upper body, and then as he rocked backward his steps would slow. The effect was like a small freighter heading through heavy seas, hurtling down from the crest of a wave and then slowly climbing the other side. He was famous for three things: his acerbic wit (he once dismissed a colleague's theory with the snort "That's not even wrong," meaning it was simply irrelevant); the Pauli Exclusion Principle, which stated that no two particles could have identical quantum numbers and with which the periodic table finally made sense; and the Pauli Effect, a magical curse that caused experimental apparatus to fail if he stepped into the laboratory.[2]

It took guts to postulate the existence of a particle that no one had ever seen. The last time anyone had predicted the existence of an unobservable quantity—the aether that was supposed to fill the universe and carry electromagnetic waves—it turned out not to exist at all. But these tiny hypothetical particles made so much sense that Enrico Fermi named them neutrinos ("little neutral ones") and incorporated them as a basic constituent (without mass) of his theory of beta decay. The theory postulated that if a neutrino actually interacted with an atomic nucleus it would, in a sense, reverse the flow of beta decay; instead of, for example, a neutron decaying to form a proton with the emission of a (negatively charged) electron and a neutrino, a neutrino hitting a proton would transform it into a neutron and emit its antimatter twin, a positively charged electron, called a positron. Such positron decays are always accompanied by two characteristic 0.511-MeV gamma rays, and a couple of decades later, in 1956, two physicists, Reines and Cowan, set up an experiment next to a nuclear reactor in South Carolina—a hotbed of nuclear reactions (the reactor, not South Carolina, which is a hotbed of reactionaries)—and those gamma rays, and therefore the elusive neutrino, were finally (indirectly) found.

We jump ahead just a couple of years now, to 1958, and to the room next door where Ray Davis was setting up equipment to measure the flux of neutrinos from the sun, which is powered by a series of nuclear reactions and beta decays, and which therefore must be emitting neutrinos. Ray's idea centered on argon.

Throughout the 1950s he had been measuring the faint radioactivities induced by cosmic rays in meteorites, concentrating on the isotope argon-37, which decays with a characteristic 35-day

2. It was the only thing he and I have in common.

half-life to the stable isotope chlorine-37. At the same time, he had always been fascinated by the mysterious neutrinos, and a year before the Reines-Cowan experiment he had thought that it might be possible to detect neutrinos by using his old buddy argon-37. His idea was that a neutrino hitting a chlorine-37 nucleus should initiate the reverse beta-decay reaction and form argon-37, which could then be measured with a Geiger counter.[3] He conducted a search similar to the Reines-Cowan work, setting up a vat of chlorine next to a reactor and then looking for radioactive argon. But he found nothing. The experiment was a failure. (In hindsight we know this was because the reactor was emitting antineutrinos, which do not react with chlorine. The Reines-Cowan experiment, without anyone knowing it, was sensitive to the antineutrinos. So Davis's negative result was the first evidence that there are these two kinds of neutrinos, although this was not understood at the time.)

Nothing daunted, he thought it would be exciting to try again, this time with the sun, which emits normal neutrinos from the nuclear reactions which provide its energy. The problem was that the sun is a hundred million miles away, so you couldn't set up your equipment right next to it, as Reines and Cowan had done with the South Carolina reactor, and the intrinsic sensitivity of their scintillation detector was too weak to find neutrinos from so far away.[4] But—aha! His argon-detection scheme was intrinsically more sensitive, and so if one exposed enough chlorine to the sun, he thought one might get enough radioactive argon-37 to be measured.

This would be lovely, Davis thought, because it would be, in a sense, seeing into the very core of the sun. Everyone knew, at that

3. Actually he was using a variation of the normal Geiger counter, called a proportional counter.

4. The neutrino flux would decrease with the same inverse square of the distance relationship that applies to sunlight.

time, just which nuclear reactions power the sun, but they knew this theoretically. It was impossible to actually see what was going on inside the sun, since the particles emitted in the postulated nuclear reactions traveled less than a billionth part of the sun's radius before they were absorbed. But the neutrinos came flying out as if the sun just wasn't there. If he could measure the sun's neutrinos, Davis would be getting the first direct experimental verification of our basic nuclear and stellar theories.

The first step was to calculate the rate at which neutrinos are emitted by the sun. For this he contacted Willy Fowler, the generally acknowledged authority in nuclear astrophysics, who in turn suggested a young theoretical physicist, John Bahcall (then at Caltech, later at the Institute for Advanced Study in Princeton; he died in 2005), who immediately became interested and carried out the complex calculations, taking the sun's energy source as postulated—basically the fusion of four hydrogen nuclei into one helium nucleus, accompanied by a series of beta decays. Then, knowing the sun's distance from the earth, he could calculate the flux of neutrinos arriving every second on every square centimeter of the earth. It was also possible to calculate the rate at which neutrinos would interact with chlorine atoms (instead of just passing through them).

And so finally Davis arrived at his experiment: he would assemble some perchloroethylene (basically, cleaning fluid), with a molecular structure of two carbon and four chlorine atoms (of which about half would be chlorine-37), and the sun's neutrinos would convert a few of the chlorine-37 atoms into argon-37, whose beta decays he could detect.

A few? How few?

Well, Bahcall's calculations indicated that if he used about a thousand gallons of the perchloro, he would get one radioactive argon atom per week.[5] One atom! The key to the experiment

5. Bahcall's initial calculation indicated even less, plunging them both into gloom, but a further refinement led to the possibility of actually measuring something.

was that argon is inert, and that one atom wouldn't react with anything. So he would bubble (nonradioactive) helium gas through it once every three or four weeks and hope to sweep up the few neutrino-formed argon atoms and pump them into his counter.

Finding those few atoms would be hard enough, but that wasn't the only problem. The surface of the earth is bathed with high-energy cosmic rays, which also produce argon-37 by a variety of reactions. Earth's atmosphere shields us from most but not all of these cosmic rays, and since they are infinitely more adept at producing argon-37, it was easy to calculate that their effect would swamp the expected neutrino effect.

No sooner did this problem arise in Ray's mind than the solution presented itself: neutrinos react so infrequently that nearly all of them pass right through the entire mass of the earth without noticing it was there, but cosmic rays are absorbed within tens of meters. So Ray decided to take his thousand-gallon equipment down into the bowels of the earth. He found a gold mine in South Dakota and received permission to do his experiment there.

That still didn't completely solve the problem, because the mass of the earth contains several radioactive elements that are continually producing energetic electrons and gamma rays which would imitate the argon-37 decays in Ray's counter. So he surrounded his counter with an array of other counters in an "anti-coincidence" array. If the particle counted in his central counter came from outside, it would have to pass through one of the surrounding array and would set that one off, too. The system would then reject that "coincidental count." Only counts originating inside the central counter (where he had pumped in the gas) would be counted. It was impossible to get rid of every extraneous count, but with the special counters he had designed and built for his meteorite work, each only 1.2 cm long and a third of a centimeter thick, he was able to reduce the background counting rate to one count a week, so the neutrino-induced counts ought to show up above this.

So anyhow, there I am in 1958, looking for something to do with my postdoctoral year, and Ray Davis invites me to join him in his hunt for solar neutrinos. Of course, as Ray explained, the experiment wouldn't be finished in a year or two, it would take many years to find enough neutrino-induced reactions to build a solid statistical case, but I could continue to work with him from wherever I went after Brookhaven. So I said, "Ray, you want to take basically a railroad car full of a dangerous, toxic, carcinogenic chemical down into the depths of South Dakota, and sit there for months at a time waiting for a couple of atoms to be formed, right? And even if you can find those atoms, why are you doing this? To find solar neutrinos, which we know are there because we understand beta decay theory and we understand the sun's source of energy. So there's no point to the experiment. The neutrinos have to be there, and when you find them, after years of difficult, dangerous, boring work, you'll just have found what everyone knew was there all along."

I didn't say this in those words, because Ray Davis was the sweetest, most polite man I have ever met, but I sure thought them as I politely declined to join the man. In my youthful arrogance I felt sorry for him, destined to spend years on a fruitless quest. It was particularly sad because he was such a gentle, lovely human being—a perfect foil for Ollie Schaeffer's penchant for practical jokes. For example, when Ray finally put together his equipment and packed it up in wooden boxes for shipment out to South Dakota, Ollie slipped into the shipping department and pried open the biggest box. Inside was a huge glass vacuum bulb, onto which Ollie taped a centerfold spread from Playboy, and then repacked it. Johnny Densieski, the lab technician the three of us shared, who went with Ray to set up the apparatus, later reported that they had a big welcome at the Homestake mine, with celebratory speeches about the beginning of a special experiment, and then they ceremoniously opened the crate—and exposed the naked lady. This was in the late '50s, when naked ladies were not ubiquitous in our society, and, as Johnny happily reported, Ray's face turned a bright red.

My favorite prank, however, was when I first got there and Ray was still involved with meteoritic research. He found an advertisement from a commercial firm offering a valuable specimen for sale. He wrote and specified that the meteorite had to be whole and with fusion crust intact. When he got an unqualified affirmative answer he ordered it. When it came to the receiving office Ollie picked it up for him. But before delivering it he secretly opened the package, removed the meteorite, replaced it with rusty iron filings, repacked it and innocently left it on Ray's desk. When Ray opened it he was livid. This clearly was not a whole specimen with fusion crust intact: he had been cheated, and for probably the first time in his life he lost his temper. He sat down and wrote a vitriolic letter and angrily deposited it in the department's outgoing mailbox.

A few hours later Ollie innocently inquired if he had received his meteorite sample, and Ray told him what had happened. Ollie smiled, and held out the real meteorite—whole and with fusion crust intact, as advertised. Ray could take a joke, and he appreciated this one; he laughed with Ollie, until he suddenly remembered his terrible letter. He ran to the Chemistry office, but the mail had already been picked up...

It wasn't till the end of the day, when he was leaving to go home, that Ollie gave him back the letter, which he had taken out of the mailbox before it was picked up.

Of all the Brookhaven staff, Ray was the friendliest. That summer we were attending the Gordon Conference on Nuclear and Cosmochemistry in New Hampshire. On the second or third morning, during the coffee break, he asked how I was enjoying it. "Great," I said.

He gave a soft smile and quietly suggested a game of tennis instead. Well, the lectures were fine, but they did tend to get a bit much. So we lagged behind as the break ended and everyone filed in, and then we hotfooted it over to the tennis court. We were in the middle of the first set when suddenly out of the sky came a loud whoosh-plop! and a hunk of something smashed onto the court. I looked up into the sky: empty, not even a cloud, just rampant sunshine. I looked at the thing: it had hit hard, but hadn't broken the surface of the court.

I began to giggle. Something falling out of the empty sky could only be a meteorite, and one not hard enough to break the court must be that most rare and valuable of all, a carbonaceous chondrite.

Davis was at the time setting up his neutrino apparatus and was simultaneously measuring the low-level radioactivities of argon in meteorites. The argon-37 isotope died out within months, and so only recent falls—rather than meteorites which had been stuck in museums for God knows how long—were useful, and among these there never yet had been a carbonaceous chondrite. And so I giggled, for while everyone else was stuck in a stuffy room listening to people talk about meteorites, I thought, Dr. Davis (he hadn't yet become Ray) and I would be rushing back to Brookhaven with a brand-new carbonaceous chondrite and measuring the argon activities, which would tell us—

I didn't get any further. Rushing over to the fallen mass, we saw it was a clump of tightly bound aluminum foil. And all my dreams turned to dust. It was chaff, stuff dumped from the Strategic Air Command's bombers to confuse enemy radar. There must be a B-36 up there so high it couldn't be seen, dropping the stuff as an exercise, and this batch hadn't opened and spread out as it was supposed to do.

Ah well, I had known it was too good to be true. We went back to tennis.

∞

I was not alone in my judgment of the worth of Ray's proposed experiment. Although he was a fine radiochemist, most of the people at Brookhaven thought it was a silly thing to do. The normal procedure for a tenured scientist was to think up things to do, and then just do them. In fact, when Ray first got there (coming from Monsanto, where scientists were assigned projects) he asked Dick Dodson, the department chair, what he was to work on. Dodson told him to go to the library and find something to do on his own; he was free to decide. Indeed, he was expected to make his own decisions.

But this idea of detecting solar neutrinos was obviously a big project, and so he wanted to get Dodson's approval. The chairman passed the buck, not wanting to take responsibility for what seemed to have a remote possibility of success, and told him to take it to Maurice Goldhaber, Brookhaven's director and an excellent particle physicist. Ray was the gentlest, most benign of creatures, the opposite of a political animal, but he wasn't stupid. He was aware of Goldhaber's statement that "No astrophysicist can calculate anything with sufficient precision to be of interest to any particle physicist." So he warned John Bahcall, who accompanied him to the meeting, not to mention the sun. They described the experiment solely as a particle study, in terms of neutrino properties, and on that basis alone it gained approval.

But it proved to be a long and arduous process, and results came slowly; at first, not at all. It was also an experiment far removed from the interests of the other members of the chemistry department. Ray became an outsider, tolerated because he had tenure, but basically ignored. And little by little the passive ignoring grew worse: the powers that be took away from him a slice of space here, a drop in the budget there, because other staff members had more important experiments to do and more immediate needs. Finally they began to push him, with hints subtle and not so subtle, to give up his tenure and look elsewhere.

By 1961 I had left Brookhaven for Cornell, and got on with my life, and so did Ray. In 1964 he finally published his first result. He had found a few decaying argon atoms which established that neutrinos were indeed coming from the sun. This was the first direct experimental evidence for the nuclear reactions that power the sun, and occasioned a loud ho-hum from the scientific community, since there really was no doubt in anyone's mind about what was going on inside our solar furnace. But Ray's physicist collaborator, John Bahcall, was able to use Ray's results to set an upper limit to the sun's nuclear furnace of 20 million degrees.

This was still not earth-shattering news, since this upper limit was well above the theoretical estimates and so didn't change anything, but it indicated that perhaps there was some use to this kind of experiment after all. Davis proposed increasing the amount of perchloroethylene to 100,000 gallons, giving an expected solar neutrino capture rate of about one per day and allowing a firm analysis of the neutrino flux. It began to look as if he might get something useful, and with the publication of his first results in *Physical Review Letters*—the nation's premier journal of physics—the honchos in the Chemistry Department at Brookhaven, while not exactly enthusiastic, began to hedge their bets and back off a little, promoting him to senior scientist.

Nobody expected anything of real importance, but people began to realize that they had been taking their ideas of the sun's furnace and what had come to be called the Standard Model of nuclear structure very seriously without experimental verification. Nothing a theoretical physicist likes better than experimental verification of his ideas (unless perhaps it's experimental refutation of some else's ideas).

And so the years passed, but as the data piled up, a curious thing became apparent: the decays expected from Bahcall's tightly drawn estimate of the number of neutrinos coming out of the sun and their effect on Davis's chlorine atoms just weren't there. Month after month, year after year, decade after decade the data increased, but the number of argon decays wasn't reaching the expected number: Bahcall's predicted neutrino flux was several times higher than Davis's measurement indicated. Each day 10^{16} neutrinos (ten million billion!) should have been passing through the tank, and just one of those should react and initiate the argon decay process. But instead of counting one a day, Davis was seeing just two per week. The discrepancy now attracted the attention of the physics community, which dubbed it the "solar neutrino problem" and soon decided it had to be a case of experimental error. Davis was, after all, just a geochemist. Who knew what errors a scientist like that could make?

But as the data dribbled in and the statistics built up, the problem, instead of going away, began to take on serious dimensions, and slowly people began to pay attention: what was going on here? Davis was invited to speak at neutrino conferences every year, and as year after year he patiently explained his rigorous attention to experimental detail he was taken more and more seriously. Bahcall's theoretical estimates of the solar neutrino flux were based on what was universally accepted as the firmest basis physics can give, both in the descriptions of the solar interior and the mechanics of beta decay. It had been vetted by other physicists, and no error was found. So where were the missing neutrinos? The Brookhaven staff began to get worried. If a simple experimental error was finally found in Davis's work, if the missing neutrinos were there after all, Brookhaven—and especially the chemistry department—would be the laughing stock of science. So Dodson took the unprecedented step of sending two senior radiochemists out to South Dakota to check out the experiment independently. Gert Friedlander and Morrie Perlman could find nothing wrong, and finally Brookhaven gave Davis its unmitigated backing, even issuing a press release about his work.[6]

By the 1970s, it had become impossible to lean on the crutch of experimental error; Davis was a consummate experimentalist and his data were beyond reproach. Something was

6. Understandably, the Brookhaven authorities have a different history in mind today. Gert Friedlander, in particular, claims that Dodson and the others always gave Davis full cooperation. Memory, theirs and mine, is a fitful thing, but here I stand to say what I do know, for I can do no other.

seriously wrong with the Standard Model of physical theory. Other people in other countries began to set up massive, expensive, more sophisticated experiments to focus on neutrinos from reactions in the sun other than those Ray was studying; by the 1990s, three other experiments were underway, in Russia, Europe, and Japan, and in all these the results were the same: the sun was either not producing as many neutrinos as it should, or something was happening to the neutrinos on their seven-minute journey to the earth. By this time it was realized that there were different kinds of neutrinos, called "flavors," and the experiments were looking only at electron neutrinos. Was it possible that these were morphing into tau or mu neutrinos—particles that would evade the detecting apparatus—during their flight?

Yes, it was possible, but only if the neutrinos had mass. Massless particles such as the photon, and such as the Standard Model of particle physics assumed neutrinos to be, cannot change their identity. (In order to do so, energy is required, and energy can be obtained by conversion of mass; if there is no mass, the required reservoir of energy does not exist.) A supernova burst in 1987 produced a burst of neutrinos; it was argued by some (and still is) that the spread in arrival times indicated neutrino mass, since massless neutrinos would all travel at the same speed (the speed of light), but this argument neglects the probability of a finite spread in time during which they were produced. In the early '90s, John Bahcall was theorizing that Davis's results could be explained only if the solar neutrinos had a tiny but definite mass, and by early 1998, nearly forty years after Ray first suggested his experiment, the three international experiments were "indicating" results that favored a neutrino mass, but, as reported in a major physics textbook, "many scientists are [still] skeptical about them." And then, later that same year, research at Japan's Super-Kamiokande neutrino detector determined that the neutrinos do indeed flavor-oscillate, and therefore have mass. Today, several other experiments, utilizing intense man-made neutrino beams as well as the sun, are actively trying to measure the incredibly small but now definitely finite mass.

So why is this of such significance? What is so important about a neutrino having an infinitesimal mass rather than a zero mass? Well, in the words of Hans Bethe, dean of theoretical nuclear physicists for at least the last half century, the neutrino data "requires new physics," the details of which can be found in the Bahcall references listed in the Notes.

To sum up, our theory of the sun's internal workings has been saved, at the cost of substantial modification of nuclear theory, and Ray Davis's stubborn fight was recognized with the award of the 2002 Nobel Prize in physics. As happens not infrequently in science, his first failure was his first success: by not finding any argon decays next to a nuclear reactor, he actually showed—though no one understood this at the time—that the antineutrino exists as a separate entity. But his big work was the solar neutrino experiment. First, by finding some solar neutrinos he provided the first hard experimental evidence for the fusion reactions that, before him, were only surmised to power the sun. In the words of John Bahcall, "This closes experimentally the scientific debate on the age of the earth and the energy generation of the sun that originated with Helm[h]oltz and Darwin in the middle of the 19th century. His experiment also created the first solar neutrino problem, the discrepancy between his measured rate and the standard model predictions." As a result of all this, he opened the door to today's energetically bursting field of neutrino astronomy, with dedicated neutrino telescopes being built around the world to study gamma ray bursts, active galactic nuclei, and other aspects of the high-energy cosmos visible to us by no other means. We are even going full circle. The first successful detection of neutrinos took place at the Savannah River nuclear plant when Reines and Cowan found neutrinos from uranium radioactivity. Today, two underground detectors in Italy and Japan are looking at neutrinos produced from the radioactivity of uranium, thorium, and potassium deep inside the earth. These geoneutrinos will reveal the concentrations and distribution of the heat-producing elements that drive plate tectonics, volcanism, mantle convection, and perhaps the core of the earth's dynamo—information desperately needed if we are to understand the current state and evolution of the earth.[7]

So the names of Ray Davis and John Bahcall go down in science history, while those of their doubters (Dodson, Fisher, et al.) are destined to be forgotten. And it all began with one or two argon atoms in a hundred thousand gallons of cleaning fluid.

7. Some thirty years ago, I tried to get such information by measuring the ^{4}He (from uranium and thorium decay) and the ^{40}Ar (from potassium decay) distributions in MORB, but the interpretation was too complex to be useful.

NINETEEN

Life and Death on Mars and Earth

The most extraordinary event of the year is the proof afforded by astronomical observations that conscious, intelligent human life exists upon the planet Mars.

—*Wall Street Journal*, Dec. 28, 1907

It's certainly not intuitively obvious that the noble gases, which don't do anything, which don't react either chemically or biologically, should be intimately related with questions of life and death, but so they are. The conventional wisdom is that they are physiologically inert: you breathe them in, you breathe them out; they don't react with your body at all. This is not quite true.

In 1938, two scientists at the Navy Experimental Diving Unit, Albert R. Behnke and Oscar D. Yarbrough, were working on the "remarkable stupefaction and neuromuscular impairment experienced by deep sea divers at depths below 100 feet" due to nitrogen narcosis. They discovered that substituting helium for atmospheric nitrogen in the air supply minimized the problem. The helium didn't actually interact physiologically, but it improved the flow of oxygen through the aerobic pathways, according to Graham's Law, which relates air flow to the inverse square root of the gas's density (i.e., atomic mass). It seemed natural to investigate the role of argon, and the following year they found that, as expected, argon (being more dense than air) interfered with air flow rather than improving it. But further experiments showed, to their surprise, that argon was even more narcotic than nitrogen, an effect they ascribed to its increased solubility in both water and fat; that is, the effect on the central nervous system seemed to depend not so much on the identity of the gas but on its total dissolved concentration in the body and blood stream.

Interesting, but soon there was a war on, and it wasn't until 1946 that several workers at the University of California, Berkeley, took the obvious next step: the solubility of the noble gases increases with atomic number, so xenon should be most narcotic. And they found this to be so; indeed, xenon was so narcotic it worked as a general anesthetic on mice. Well, that *was* interesting. The whole idea of anesthesia was still rather new—I was a teenager at the time, and dreaded the twice-yearly visit to my dentist because (without fluoride) I always had a few cavities, and my dentist still didn't believe in Novocain—and it took another five years before anyone tried it on people. It was in 1951 that the medical profession chipped in, as two investigators from the departments of anesthesia and surgery at the University of Iowa, Stuart Cullen and Erwin Gross, found not only a decreased sensitivity to pain and eventual loss of consciousness when volunteers breathed a 50–50 mixture of xenon and air, but "a pronounced narcotic effect" with an 80–20 mixture. Having the courage of their convictions (and hopefully, in those days of limited patient rights, informing and convincing their patients), they performed two operations using xenon as the anesthetic, and it all went beautifully. Indeed, xenon turns out to be ideal for this purpose: "Inhalation induction is smooth and rapid, with rapid emergence seemingly regardless of duration of anesthesia."

On the other hand, it costs more than the usual anesthetics, and so interest has lagged. It wasn't until the 1990s that the establishment in Europe and Japan began to take note; by 2000 it was being used in Russia, and by 2005 in Germany. It is still being ignored in America. It is being used, however, for another purpose in its radioactive form. If you've been unfortunate enough to have a serious pulmonary problem, you might have been fortunate enough to live in a community with a well-staffed nuclear medicine facility. If so, you might have been given some xenon-133 to breathe.

The problem we're talking about, called a pulmonary embolism, is a fairly common and sometimes fatal condition that arises when a blood clot in a vein, usually in the legs or pelvis, breaks off and travels with the blood through the vein. It moves through progressively larger veins until it reaches the relatively cavernous chambers of the heart. In the heart, it might get chopped up a bit into smaller pieces, but the clot or clots continue moving with the blood through the large pulmonary arteries into the lungs. There, they encounter smaller arteries until they can pass no further. They get stuck, and clog the arteries they are in. Blood flow to part of the lung is suddenly obstructed, causing a reduction in oxygenation of the blood in these areas. Typically, the patient rapidly becomes short of breath and often experiences sharp chest pain. The condition can mimic a heart attack, but can also be difficult to distinguish from other pulmonary conditions such as pneumonia, aspiration of food, or an acute asthmatic attack, just to name a few.

In the late 1960s, an imaging test was developed to help identify pulmonary embolism. Patients were injected in an arm vein with a large number of very tiny protein particles which were bound to radioactive Tc-99m.[1] The particles traveled with the blood to the lungs in the same manner as a blood clot, but being much smaller, they spread diffusely into the tiny capillaries in the lungs, where they got stuck. (Since they filled only about 0.1% of all the pulmonary capillaries, they didn't cause any problems, nor does the small amount of radioactivity.) The Tc-99m gamma rays allowed their distribution in the lungs to be captured as an image using a gamma camera.

1. A metastable state of technetium, which emits gamma rays in decaying to the ground state.

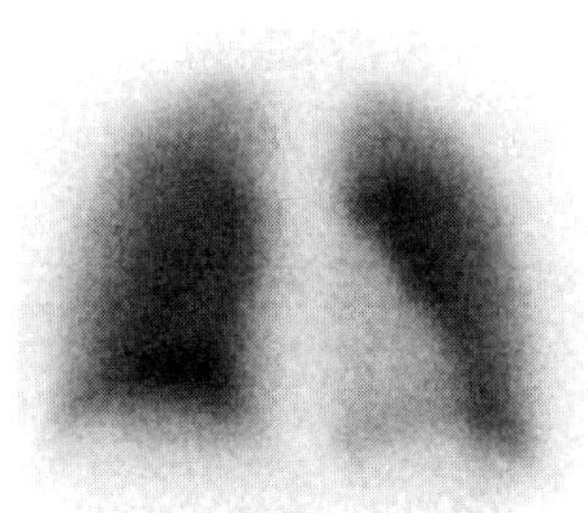
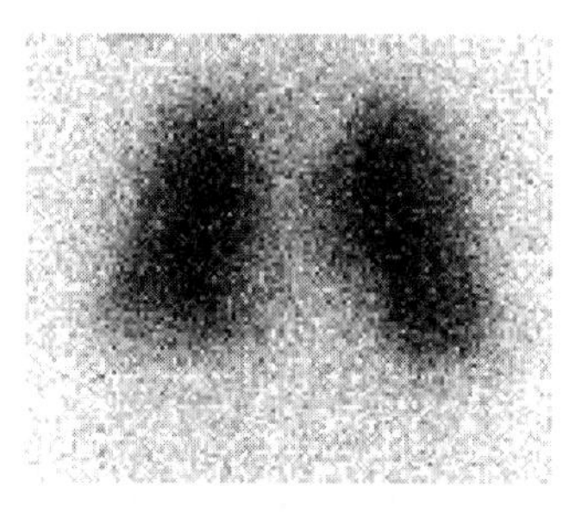

Normal VQ scan. *Left*: Perfusion with Tc^{99m}-protein particles, injected intravenously. The lung perfusion is in black. The white areas include the space between the lungs, and the silhouette of the heart. *Right*: Ventilation with Xe^{133}. Ventilation and perfusion images are normal and similar.

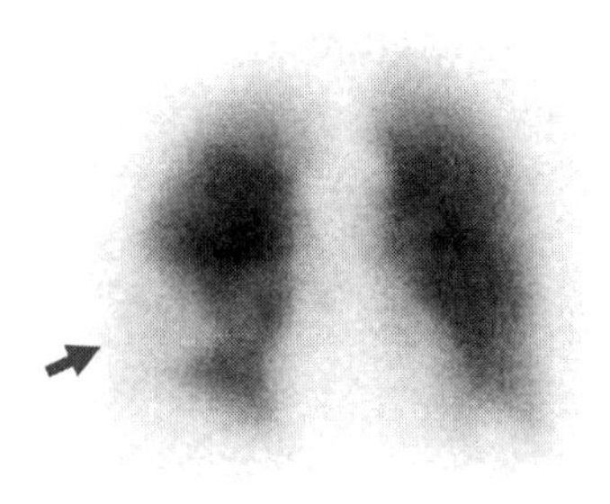
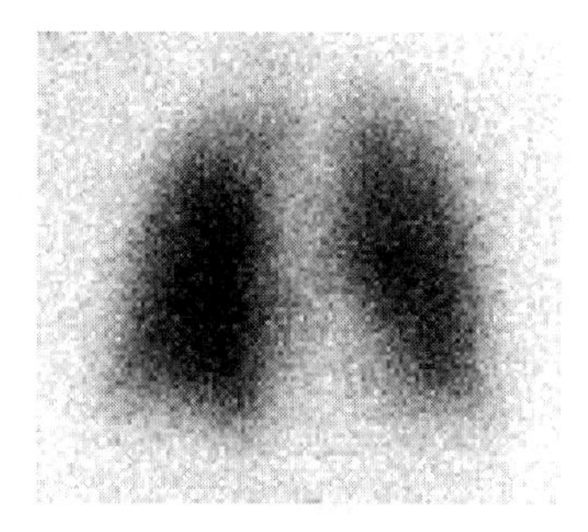

Pulmonary Embolism. *Left*: Perfusion images with Tc^{99m} show a large defect in the lower right lung (arrow). *Right*: Ventilation images with Xe^{133} show normal ventilation in the region of the perfusion defect, indicating that the perfusion defect is due to pulmonary embolism.

FIGURE 7. ^{133}Xe (Ron Fisher, MD, PhD, The Methodist Hospital, Baylor College of Medicine, Houston, Texas)

The problem was that *any* pathological process in the lungs (pneumonia, asthma, etc.) produced areas of reduced blood flow that were difficult to distinguish from an embolism. To correctly diagnose the problem, a procedure was needed that could isolate and demonstrate the pattern of air flow in the lungs. What was needed was an inert but radioactive gas. Enter xenon; in particular the radioactive isotope 133, which also emits gamma rays. The patient breathes xenon-133 from a mask connected by tubing to a closed ventilation system. (Because of the sensitivity of radioactivity measurements, only a small amount of xenon is needed, not enough to induce narcosis.) Within a few minutes, a series of images are obtained that show how the gas distributes in the lungs. In the case of pulmonary embolism, the ventilation images are normal, but the blood flow pattern is not (figure 7).

The test has been so successful that literally millions have been performed over the past four decades. Today, the pulmonary arteries can be imaged directly with CT angiography, but the xenon scan continues to be used for the ~20% of patients who are not candidates for CT scanning because of other physical problems.

Meanwhile, the gases have been involved in self-assisted suicide, life on Mars, the origin of life and the future of civilization on earth, and the search for Hitler's atomic bomb. Let's begin with the last, since it also involves xenon-133, and work our way backward.

One night after dinner in the 1960s I was talking with Phil Morrison, who had spent the war years working on the Manhattan Project. He said that by 1944 it was becoming clear that it would be possible to make an atomic bomb, so by his bedside in Los Alamos he had kept a short-wave radio tuned to London. Every morning he would wake up and turn it on, then turn it off right away and get up and go to work. He wasn't interested in what London was saying, he just wanted to know that London was still there—that there wouldn't be just static on the BBC, that Hitler's scientists hadn't beaten us to the bomb.

It wasn't an irrational fear. After all, it was in Otto Hahn's chemistry laboratory in Berlin that fission had first been recognized.[2] And although the bulk of Germany's scientists had fled Hitlerism—people like Lise Meitner—many good ones remained, including Hahn and Werner Heisenberg, one of the world's top theoreticians. So Morrison's fear was well-founded and prevalent around the Manhattan Project, until finally the commanding general, Leslie Groves, called in an experimental physicist to talk about it.

Hypothetically, he said, if the Germans were building the bomb, how could we find out about it?

I'll have to give that some thought, the physicist, Louis Alvarez said.

Do that, General Groves said. Report back in one week.

And so he did. He focused on the fission products: radioactive nuclides that form when the uranium nucleus is split. One of them in particular stood out: xenon-133. Being a noble gas it wouldn't react with anything and so would inevitably leak out of the reactor and escape into the air. With a five-day half-life it would hang around long enough to be detected, its gamma radiation would be easy to identify, and its presence would be definitive evidence of uranium fission, that is, a working nuclear reactor. When/if found, the B-24s and B-17s of the Eighth Air Force would have a vital new mission.

Alvarez designed a filter trap in the belly of an A-26 Invader attack bomber that would fly low over possible German sites and sniff for xenon. Actually it would pick up a variety of gases, but taking advantage of the non-chemistry of the noble gases they could

2. Fermi had first split the uranium atom in Rome, but didn't know it. Hahn thought it had been an error, but Lise Meitner (a theoretical physicist) nagged him to repeat the experiment, and it was she who recognized the result and named it fission.

be cleaned up (as in a mass spectrometer) to leave only xenon, which would then be pumped into a counter for identification.

The equipment was loaded and the A-26 flew over suspected German sites, and detected nothing. This was the first indication that the Nazis had made no significant progress toward a bomb, but it brought little comfort. Morrison kept his radio tuned to London every morning, for the experiment had an inherent weakness, one to be avoided whenever possible but unfortunately common to many. A positive result would have been absolute confirmation of what they were seeking, but for several reasons a negative result, though reassuring, was not absolute proof. Xenon is a gas, but much heavier than air. So when released to the atmosphere it doesn't float upwards but instead sinks to the ground, and how low can an Invader fly? Additionally, the sensitivity of the apparatus was poorly known, its range was questionable, the suspected German atomic sites might have been entirely the wrong sites . . . etc., etc.

Oh, well, as Emiliani said. Nobody is perfect.

Next on our list is the origin of life and the future of civilization on earth. About a decade after I left Cornell, Tommy Gold came to Miami for a visit, and after a few beers he asked me what I knew about the relationship between helium and petroleum. I shrugged. "Shouldn't be any relationship," I said, and he smiled and nodded.

"Shouldn't be," he agreed. "But there is."

This was about five years after the first lunar samples had been brought back by Apollo. The frenzy about analyzing them was still going strong, and I couldn't imagine why Tommy was thinking about such terrestrial stuff as petroleum. The moon had always been one of his consuming interests; he had wanted to hire me into his Center for Radiophysics and Space Research to do activation analysis on the samples, and his had been a lonely and scorned voice arguing that the surface of the moon would be powdery dust instead of hard rock. But then I remembered that none of his consuming interests entirely consumed him; he was always thinking of something else. In 1968 it had been the newly discovered pulsars; today it was helium and petroleum.

The pulsar story was illustrative of why he should always be taken seriously. In 1967, a graduate student at the University of Cambridge, Jocelyn Bell (now Burnell), discovered the first one, a star that appeared to be pulsing its energy so regularly (every 1.333 seconds) that the radiophysics staff named it, only half jokingly, LGM (Little Green Men), thinking it could be a signal put out by an alien civilization. So many other pulsing stars were soon found that the interpretation was discarded in favor of something natural—"something," but no one knew what, for it was impossible for an object the size of a star to pulse so coherently on the short time scale observed.

Gold almost immediately hit on the solution: the pulsar was actually the neutron star hypothesized back in 1934 to be the end result of a supernova explosion. Thirty-something years later, it had still never been observed. Gold pointed out that conservation of both angular momentum and magnetic field when a normal star condensed to only a few miles across would result in its spinning rapidly and emitting radiation along its magnetic pole axis. If the magnetic axis was different from the rotational axis, the star would look just like a rotating beacon: you would see its light as a "pulse" when it swept across your line of vision.

If I'm right, he said (and if neutron stars truly are the result of supernova explosions), we should see pulsars where once we saw supernovae. Nonsense, said the astronomers. The pulsar phenomenon was so surprising that a conference was immediately called to discuss it, and explanations were asked for. But when Tommy submitted his paper, it was rejected as too fantastic. Sounder heads prevailed at *Nature*, where it was subsequently published, and the next year when a pulsar was discovered at the center of the Crab Nebula (which is a remnant of a supernova spotted by the Chinese in 1054), and when other pulsars were observed to slow down their spin as he had predicted, the explanation was accepted.[3]

So when he started talking about helium and petroleum, I listened.[4]

The accepted explanation for the origin of petroleum is, as every schoolboy knows, that it's a "fossil fuel," formed by the heat- and pressure-assisted decomposition of ancient life forms. (All living creatures, from asparagus to zebras, are formed of organic molecules, primarily carbon and hydrogen, known as hydrocarbons; petroleum—whether gas or oil—is made of the same stuff.) This explanation is as well founded and as universally accepted as that for the "aether" which at one time was understood to fill the universe. (Remember Lord Kelvin? "If there is one thing we are sure of concerning the universe, it is the existence of the aether.") Of course, there are many more universally accepted explanations of scientific phenomena which have *not* turned out to be wrong; by the time a consensus has been reached among scientists, the explanation is probably correct. Still ...?

The story Tommy tells is frankly exasperating to nearly all petroleum petrologists, just as his explanation of a powdery surface of the moon was to NASA, his explanation of hearing was to audiologists,[5] and his explanation of the pulsar was to astronomers. Like those stories, this one about the origin of petroleum is interesting and provocative, and perhaps, like those, it may turn out to be right.

It begins, as he told me that day, with the unexplained but silently accepted association of helium with the hydrocarbons

3. When a Nobel Prize was awarded, it didn't go to Jo Bell, who discovered the first pulsar, or to Gold, who explained them, but to Anthony Hewish, the head of Bell's lab, who had initially rejected both Bell's observation and Gold's explanation.

4. On the other hand, a lot of us listened when he attempted to demolish the Big Bang with his innovative Steady State theory of the non-origin of the universe, a theory which though quite lovely turned out to be quite wrong.

5. He also overturned the classical (Helmholtz, nineteenth century) theory of how our ears hear; his idea wasn't accepted for thirty years, but is now known to be right.

that form petroleum. The data themselves are incontrovertible. When the U.S. Bureau of Mines wanted to find commercially important helium sources, the places they looked were known gas and oil fields, and there they found the helium. And conversely, every petrologist knows that where you find helium, you're likely to be looking at a petroleum-rich field. Portable helium detectors, known as "snifters," are routinely used in petroleum exploration, and where you find a lot of helium seeping out of the ground, it turns out to be a good bet that there's some exploitable petroleum hiding under it, so start drilling and don't ask questions.

Surprisingly, despite the very strong association, aside from a few Russian geologists who were not part of the Western discourse no one asked the obvious question until Gold did: "Why should a chemically inert gas like helium be associated so strongly with ancient biological remains?" Even now, two recent reviews go into great detail about the monumental sets of data, but never once address that question.

But perhaps it's not surprising. To challenge—even to politely question—the hard-core belief of generations of petrologists is a daunting task. What takes it into the realm of sheer entertainment is the way Tommy Gold went about it: not politely questioning, but arrogantly informing all of them that they were not only wrong but damned fools for not seeing the obvious.

Well, that was the way of him, God rest him. I remember one day at a filled conference hall when his breezy dismissal of the idea of God (a laughable concept, he called it) brought at least one Catholic scientist, John O'Keefe, to his feet in a sputtering, stuttering, red-faced frenzy. In much the same manner, instead of begging their pardon for suggesting a heresy, he simply informed the petrologists that the petroleum hydrocarbons are not fossil remains at all, not formed over the aeons by the pressurized and heated decay of once-living organisms, but are instead a primordial goo left over from the formation of the earth. How else to explain their close association with helium?

What he told me that day he has continued to hammer home: "The association of helium with hydrocarbons is probably the most striking fact that the biogenic theory fails to account for."

The standard response of the traditional petrologists is "Ho hum, whatever." Just a minor unexplained point only slightly marring a complete theory. For if petroleum is not a fossil fuel, they ask, why is it found precisely where traditional petrologists look for it (based on their fossil fuel model)?

Easily explained, Gold laughs. Because it's everywhere. So no matter how false your theory is, if it tells you to look someplace, that's where you'll find it. You think success verifies your theory, but any theory will give the same result. Actually it's buried deep in the mantle and distributed everywhere on earth. Dig deep enough at any random spot and you'll find it.

Ah, "deep enough"; there's the rub. How deep is deep enough? That depends on the overlying rock. Gold's thesis is—well, first, the conventional theory: since the various components of petroleum are a series of organic molecules, hydrocarbons, which are the chemicals of which life is composed; and since wherever petroleum deposits are found, they are intimately mixed with various biological markers—that is, clear indications of living processes—it clearly follows that the petroleum is the remains (the fossil remains) of living organisms.

No, says Gold. First, the undisputed fact that petroleum is composed of organic molecules is not prima facie evidence of past life, for such organics are found in conditions where no one pretends life is to be found: in meteorites, in vast seas on Titan, in comets and asteroids, floating around in space[6]—in fact, everywhere. Organics are ubiquitous in the universe. They are the building blocks of life, yes, but not necessarily the products. They

6. On April 21, 2009, workers in Germany reported at a meeting celebrating the European Week of Astronomy and Space Science that they had identified the presence of ethyl formate and n-propyl cyanide in interstellar space. Such complex organics had previously been thought to be biogenic, but there is no life in interstellar space.

are necessary but not sufficient evidence of life: you can't have life without them, but you can certainly have them without life.[7]

Okay, everyone agrees with that. It's the second point of his thesis that irritates: while biological markers are always found in petroleum fields, it's not because the hydrocarbons there are the remains of life but instead because they are good food for life. Wherever petroleum is, life will follow in its endless search for food. The deep earth is literally crawling with life, he insists: it's what he calls the deep hot biosphere. It follows from these two points that since organic compounds form naturally and abundantly in the universe, they would have formed a substantial part of the particles that formed the earth, and therefore we can expect the deep interior of the earth to be rich in them. Also distributed throughout the earth—with no possible causal relationship to living creatures or their fossil remains—are uranium and thorium, which radioactively decay to produce helium. So how does the petroleum come to be associated with helium?

Couldn't be simpler, Gold explains. The helium is produced as individual atoms, which are generally trapped in the rock matrix deep underground as well as in the upper tier of sedimentary rock (where the biogenic theory says all petroleum deposits must be). The diffusion of helium through the rock strata toward the surface is a very slow process. Petroleum deposits, on the other hand, are present not as individual atoms but as vast underground fields, and as such they can break through weak rock structures and force their way toward the surface. As they do, Gold reasons, they will sweep up any helium atoms they meet along the way, just like the helium Ray Davis used in order to flush out the individual atoms of argon formed by neutrino reactions in his tank cars of chlorine; the gaseous hydrocarbons—mainly methane—will act as the same sort of carrier gas, in this case bringing the

7. Glycine, an amino acid used by living organisms to make proteins, has recently been found in a comet. On earth, its presence would certainly be claimed to be a biomarker, indicative of life, but no one pretends there are living creatures in a comet.

helium along wherever the hydrocarbon concentration is great enough. In fact, the amounts of helium present will be an indication of the depth of origin of the methane, for the longer the pathway to the surface, the greater the amount of helium brought along. His analysis shows that helium comes from the deepest levels, gas from the next deepest, and oil from the next, but all of these come from levels much deeper than the crustal sedimentary layer where the biogenic theory says all petroleum must lie. This is buttressed by an independent report of helium and methane in Canada: "The coincidence of methane and helium anomalies with known tectonic features also indicates fracture leakage from depth and the possible existence of oil and gas fields"—that is, deeper in the earth than the sedimentary layer postulated by the biogenic theory.

If Gold is right, the earth holds vast reservoirs of petroleum in general and methane gas in particular. The deposits we have found are only the tip of the iceberg, where the local geology allowed the stuff to percolate close to the surface. If we could just dig deep enough, we'd find it everywhere. But deep enough means miles deep, beyond our current capabilities.

There was one serious attempt. In 1983, Gold persuaded the Swedish State Power Board to put up the money and drill through the Siljan meteorite impact structure which had shattered the ground to rubble in the central part of the country, providing the perfect conditions for deep mantle methane to percolate upwards to a reasonable depth. He calculated that they would hit a pocket of methane at seven to eight kilometers. Unfortunately, at 6.7 km they ran into an oily sludge that clogged the drill, and they had to stop drilling. They did bring up eighty-four barrels of oil and detected slightly more than a whiff of methane at the surface, enough for Gold to declare a scientific success but not enough for everyone else to agree.

The consensus remains strongly against him, but new trickles of data keep popping up, teasingly vague but persistent. In the July 26, 2009, issue of *Nature Geoscience*, an article titled

"Methane-Derived Hydrocarbons Produced under Upper-Mantle Conditions" states that "hydrocarbons heavier than methane can be produced by abiogenic processes in the upper mantle." And in 2010 further evidence by the same group was published in *Reviews of Geophysics,* arguing that "Experimental results and geological investigations presented in this article convincingly confirm the main postulates of the [abiogenic] theory."

If it turns out that Gold is right, the consequences are twofold. First, the quantities of methane in the earth as primordial relics would be orders of magnitude more than methane as biological remains, totally changing the character of our dependence on "fossil" fuels; there would be enough to go around for everyone for the foreseeable future, and Saudi Arabia could go suck lemons.[8] Second, the abundance of heavier organic molecules, sludge, and oil associated with methane would be the feces of bacterial life deep in the earth: the deep hot biosphere. Gold conjectures that it is with these subterranean forms that life first started, arguing that the acquisition of energy—the first step toward living systems—would be much simpler in the oxidation of methane than in photosynthesis.

This step, the progression from organic molecules to a system of extracting energy from the environment and using it to grow and evolve, is the creation of life and is the biggest mystery in the universe today. We simply don't know how it happened. The most important data bearing on the subject—data which are still completely missing—are any sort of indications of life somewhere else in the universe. Anywhere else, any kind of life.

Up to now, the attempts to find extraterrestrial life have focused on surface life, based on our understanding that solar energy is necessary, and there is nowhere else in the solar system where conditions are favorable for surface life. But if Gold is

8. This would still leave us with the problem of greenhouse gases.

right and life thrives internally, the stage opens up considerably.[9] Provocative evidence from the Siljan hole was the presence of microscopic grains of magnetite in the oily sludge, grains similar to those produced by bacteria on earth, a possible indication of life deep, deep down.

Which brings us to the next impact of the noble gases: life on Mars.

On July 20, 1976, a sophisticated scientific package extricated itself from a Viking rocket and landed on Mars with one object in mind: find evidence of life, or definitive evidence of its lack. Eleven days later, the *New York Times* reported . . .

> July 31: "The first results look very much like a biological signal."
> August 3: "The behavior is *not* characteristic of a biological reaction."
> August 8: "Tests by Viking strengthen hint of life on Mars."
> August 11: "We may not be able to answer that question with Viking."

For the next twenty years, the argument went on, with most scientists agreeing that the question wouldn't be settled until we actually brought back a piece of the planet. Then, on August 7, 1996, NASA called a special press conference to announce to a startled world the publication of a research paper detailing a new investigation of the planet Mars, the concluding sentence of which stated: "We conclude that they [the data] are evidence for primitive life on early Mars."

The announcement took everyone by surprise. We hadn't sent any spacecraft to Mars since Viking. What had happened?

9. A related possibility is life in internal oceans, with reference to the Jovian moons Encephalus and Europa. See, for example, my book *Strangers in the Night: A Brief History of Life on Other Worlds*.

What had happened was a series of investigations that began even before Viking, in 1968, with curiosity, expanded in 1980 with a ridiculous suggestion, began in earnest in 1983, culminated some fourteen years later with NASA's stunning words, only to fade away like the Cheshire Cat in the next century, leaving behind just the trace of a smile.

What had happened, in fact, was the discovery of a poor man's sample of Mars, one that was delivered free of charge to Earth. But the wrapping was unusual...

We've talked earlier about the 4.5-billion-year solidification ages of meteorites, the time at which they became solid bodies and started retaining their radiogenic argon, taken as the time of origin of the solar system. These primitive meteorites are called chondrites, but there is another group of meteorites, the achondrites,[10] that are similar to terrestrial igneous rocks in that they have been melted at some later date. Three of them, in fact, were found in 1968 to have melted at much too late a date. The data indicate that they melted less than a billion years ago.

This was impossible, or at least curious, because the body or bodies in which they melted could *not* have melted at such a late date. Planetary bodies are heated on the inside by radioactive elements, mainly potassium, thorium, and uranium, and by residual heat left over from their formation, and they lose heat by radiating it away. The volume of such a body is proportional to the cube of its radius, while its surface area is proportional only to its square, so the larger the body, the greater the ratio of its volume to its surface area. Heat is generated in the volume and radiated away by the surface area, so smaller bodies lose heat more easily; in other words, the smaller the body, the less radioactivity and the higher the surface-to-volume ratio, which means both that there is less heat generated and that the heat escapes more rapidly to space. Asteroids were the generally postulated bodies of origin for

10. The chondrites are characterized by small nodules, chondrules, that were formed in the solar nebula. The achondrites are those meteorites without chondrules.

meteorites, but bodies as small as asteroids can't generate enough internal heat for volcanic activity. Nor could they have come from the moon; volcanic activity there ceased nearly four billion years ago as its heat was lost to space.

So where else is there a planet large enough to have retained enough internal heat to melt stone as recently as a billion years ago? The large Jovian planets, certainly, but their gravity is so intense you couldn't possibly eject a rock from their solid interiors (if they have one). Mercury, like the moon, is too small. That leaves only Venus and Mars. For dynamic reasons, Venus doesn't look likely (its gravity and thick atmosphere make it difficult to conceive of an asteroidal collision knocking large chunks off it and out into space). In 1980, two graduate students, Hap McSweeny and Ed Stolper, suggested that this particular meteorite must have come not from an asteroid but from a much larger body, in particular from Mars.

They were not actually the first to suggest it, but they were the first to say it out loud, so to speak. Others were sort of dancing around it; the idea was blowing in the wind. A year earlier, in a paper given at the Lunar Conference in Houston, Stolper and two others whispered, "A martian derivation for some of these meteorites cannot be ruled out." And at the yearly meeting of the Meteoritical Society, several other people put their heads together and sort of said it: "[The data] imply a sizable parent planet. A portion of the planet's surface must be young. Mars fits these criteria." And the momentum just grew until finally McSweeny and Stolper came out with it. After considering and dismissing all other sources, they wrote: "The only alternative source . . . is Mars."

An asteroid crashing into the planet, they said, must have knocked off chunks of the surface which were thrown into space and eventually fell on earth. But this suggestion was greeted with disdain by the theorists, who showed that an asteroidal collision providing enough energy to blast a piece of the planet into space would actually vaporize it; the idea was impossible.

Well, McSweeny and Stolper knew that. But data are nasty things: everyone is entitled to their own ideas, but not to their own data. And the data said that the three meteorites—known collectively as Shergottites after the first one analyzed, the Shergotty meteorite—melted inside a body as large as Mars. Theoreticians, though, can be as nasty as data, or even nastier, and they all scoffed[11]—until 1983, when two scientists at the NASA Johnson Space Center in Houston picked up a fourth Shergottite, one that had been found lying on the ice in Antarctica, and measured the noble gases in it. Remembering the Mars suggestion, they compared their results to the gases found in the Martian atmosphere. The title of their paper reflected the prevailing climate: "Martian Gases in an Antarctic Meteorite?" But further data on other Shergottites[12] soon tightened up the precision and removed the question mark: the abundances of the noble gases in these meteorites are an excellent fit to the abundances in the Martian atmosphere and totally different from those on the moon or Earth, and it is now universally accepted that they come from Mars. In the face of such nasty data, the theoreticians retreated, and it was soon acknowledged that it was possible to knock pieces of the planet off into space, after all.

It was on one of these Shergottites, or SNCs, that NASA found what they considered to be five distinct evidences of life, none of them definitive by themselves but, taken together, mighty indicative.

First were tiny "critters," revealed by scanning electron microscopy, which looked remarkably like the first simple forms of life found on earth. Second, carbonate minerals that looked like those formed on earth only by precipitating from water, and water is the sine qua non for life. Third, oxidized and reduced forms of iron

11. So did I.

12. The Martian meteorites have grown to eleven in number and are now commonly referred to as SNCs (snicks), the name coming from the first three to land on earth: Shergotty, Nakhla, and Chassigny.

in the same minerals. Fourth, polycyclic aromatic hydrocarbons, which on earth are associated with living organisms, and fifth, magnetite grains.

Their paper reverberated throughout the scientific community and sparked so much interest that exobiology, the study of life beyond earth, became for the first time an established scientific discipline as centers were established in several universities and national research labs. The field continues to flourish today, although most of the original impetus has fallen by the wayside.

The microscopic critters are now thought by most workers to be too tiny to be biological; they don't have enough internal volume to hold the complex molecules of life, although some have suggested that they may be pieces of larger organisms. Carbonate minerals similar to those found in the meteorite have been produced at high temperatures without the presence of water. The occurrence of minerals both oxidized and reduced was at first thought to be impossible except for the processes of living organisms, but later research has found inorganic explanations. Similarly, although polycyclic aromatic hydrocarbons are often found on earth in conjunction with life, they can occur inorganically.

Finally, the small grains of magnetite were perhaps the most convincing evidence (remember Gold's finding in the deep earth oil sludge), but again it has now been shown that they can be produced without the presence of life. So once again we're left with a tantalizing possibility, but with no definite proof. And just when things were dying down again, guess what? Plumes of methane have been found spewing out of Mars. According to the biogenic theory, in which methane is a fossil, this indicates subterranean life on Mars, although the amounts seen would seem to indicate an improbably high abundance of such life on a planet generally accounted to be barely suitable for living organisms. On the other hand (once again), the observation could be viewed as confirmation of Gold's thesis that methane is a normal, primordial constituent of all planets, which in turn implies an abundance of subterranean life. Who knows?

Stay tuned. Or, if you prefer to check out early because you're incurably ill and in constant pain, there's one final use for helium on earth: a pleasant way to ease your path out of this life. All you need are a couple of helium tanks, the kind used to fill balloons, available at any party store, and either a hood kit (available online) or a simple plastic bag and a large rubber band to seal it. The idea is to breathe in helium instead of oxygen.[13] Now, I haven't tried this, but evidently your body can't tell the difference, and so it doesn't struggle or exhibit any sign of stress. You just quietly go to sleep.

And so, good night.

13. Any of the noble gases would do as well, but helium is easily available.

TWENTY

Radon and You

I seen a dog go in that mine that couldn't hardly walk . . .

—"Wild Bill" Remior

A REVIEW FOR ONE OF MY NATIONAL SCIENCE FOUNdation grant proposals began "He isn't the world's greatest experimentalist, but . . ."

I couldn't argue with that. In fact, once I nearly destroyed the Brookhaven chemistry building. I had been working nearly thirty-six hours straight and was a bit tired. When I finally left, with the sun not quite rising, I forgot to turn off the Bunsen burner. Not only that, but I had left it with its rubber feed tube twisted a bit. Evidently the strain on the tubing later pulled the burner around and tipped it over. When it fell, the flame went out but the gas kept right on flowing through, and a few hours later when our lab tech, Johnny Densieski (who was always the first person to come in every morning), opened the doors he was nearly knocked over by the smell. Ordinarily he would have been more than just *nearly* knocked over, but for some reason the lit cigarette that was always dangling from his lips was absent that day, saving his life and the chemistry building from instant demolition.

Johnny was loyally close-mouthed about it, but the story did get around. So when a routine exam of our lab showed radioactive contamination, everyone's eyes naturally turned to me. I have to admit I thought it probably was my fault until a spectrometric examination of the evidence showed that it was alpha radiation, and all my work was done with beta radionuclides. The health physicists conducting the investigation didn't believe me, of course, and so they tested lab counters and desks everywhere I had been, and found more and more radioactivity. They were about to put me in front of a firing squad until, being more thorough, they began to test areas where I had *never* been and found the same radioactive contamination,[1] which they identified as polonium-218.

No one, it turned out, was working with polonium, yet it was everywhere. Agatha Christie couldn't have devised a better puzzle. And just as in one of her mysteries, a final clue provided the answer. Although the contamination was everywhere in the building, it turned out to be concentrated around the air-conditioning vents.

Get it? A gas had to be the culprit, something that the air-conditioning system would spread around the lab. Once that was realized, the identity was clear. Polonium-218 is formed from the radioactive decay of the noble gas radon-222, which in turn forms from the decay of radium-226, which was one of the radioactive nuclides being studied in one of the labs. Nobody there had spilled anything and tracked it around the building, but some of the radium decayed to radon during the course of the experiment, and the radon wafted out of the beaker and into the air-conditioning system, which, though grossly inadequate to cool the lab during the Long Island summers, was efficient enough to distribute the radon around the building. Eventually, the radon decayed to polonium, which settled on the counters, desks, and floors and was finally detected during the health physicists' routine workplace examination.

Radon, the final rare gas, was both discovered and today is killing thousands of people every year in much the same way.

1. This can be seen as an illustration of Tommy Gold's criticism of the biogenic fuel theory. The health -physicists' theory—that I was the source of the contamination—predicted that it would be found wherever I worked. And it was! But the theory was wrong. (The biogenic fuel theory predicts where fossil fuels are to be found, and they are! But...)

In the 1880s, a Polish student named Maria Sklodowska got herself involved in some disreputable revolutionary activity in Warsaw and thought it advisable to leave, temporarily giving up her university ambitions. She agreed to work as a governess, sending part of her earnings to her elder sister, Bronisława, to enable her to study medicine in Paris (for in Poland, then under Russian rule, women were not allowed in university); upon receiving her medical degree, Bronisława would then pay for Maria's studies in science. In 1891 Bronisława sent for her and she finally came to Paris to live in a garret and study at the Sorbonne.

Before she finished her training, she fell in love and got married, apparently justifying the Russian concerns about professors wasting their time training women because they just quit and get married and have children and are never heard from again. But she fooled them. The man she married, Pierre Curie, was also a scientist, and the marriage detracted nothing from her scientific efforts.[2]

By 1896 she was looking for a thesis subject. By chance, that year another Parisian scientist, Henri Becquerel, discovered curious electromagnetic emanations from uranium ore, and Marie Curie (as she now was) decided on this as her subject. Becquerel's discovery was that time's equivalent of the twentieth century's "discovery" of cold fusion—a phenomenon that according to all nuclear theory could not happen, but which immediately brought in millions of dollars of research funds and no shortage of scientists jumping into the game. It is interesting that Becquerel's rays, on the contrary, created no fuss at all; since it was impossible for a rock to generate energy, it couldn't happen, and aside from Mme. Curie (and soon, her husband) scientists simply ignored it. Sometimes the times do change.

By 1898, just two years after Becquerel's discovery, she and her husband had chemically isolated two new elements—radium and

2. Her daughter, Irene Joliot-Curie, would also win a Nobel Prize in chemistry. And *her* daughter (and son) are both scientists.

polonium (named in honor of her homeland)—from Becquerel's uranium mineral pitchblende, obtaining with infinite patience and a lot of hard work just a few grams of each from several tons of the ore, and had pinned down the properties of the peculiar phenomenon (which she named radioactivity) so definitively that everyone who had doubted its existence was convinced. Five years later they shared the physics Nobel Prize with Becquerel, and in 1911 Madame Curie won the chemistry prize all by herself, becoming the first person to win two Nobels in science.[3]

In the course of her studies she noted the appearance of a radioactive gas that rose out of her beakers when pitchblende was dissolved, but she didn't follow this up except to note that it disappeared in about a month, and it was a German scientist, Friedrich Ernst Dorn, who first isolated it and named it radium emanation (Ra Em). A year later Rutherford found a similar gas bubbling out of thorium compounds, except that it disappeared within minutes. The spectra of these two emanations, and those of another gas appearing from actinium, were identical to each other and similar to those of argon, krypton, and xenon, so in 1904 Ramsay suggested that they were different forms (now called isotopes) of a new rare gas element, finally known as radon.

And now fifty years flutter by...

...and it's 1954 at the General Electric research laboratory in Schenectady, New York. In those days (now, unfortunately, long gone), it was considered reasonable for a profit-making company to support research that might not have immediate commercial value, and so two of the most prestigious research labs in the United States were those of General Electric and Bell Telephone (in New Jersey). Today this idea has been scrapped, on the

3. There have only been two others since then.

theory that commercial applications must be in mass production by tomorrow at the latest or the research isn't worth doing. (Enter Japan, laughing...)

But now we're back in 1954 and Bob Walker, upon obtaining his PhD in particle physics, has joined the GE staff because the Korean War is raging and his draft board threatened to draft him if he did as he originally intended, begin a career as a university professor, but said they would leave him alone if he took a research position in a laboratory necessary to the war effort. Happily, in Schenectady he found that he was not required to do any war work but was free to follow his scientific instincts so long as they showed any possible future correlation with GE's destiny. He began a study of radiation effects, which was to split asunder into a host of different directions.

It began a few years later, in September of 1960, at a scientific congress in France, where Walker met the British scientist R. S. Barnes, who showed him a few pictures taken with an electron microscope of tracks left in mica when it was irradiated with fission fragments. It was just an amusing little effect to Barnes, of no real use because the tracks faded under the electron beam within seconds, but to Walker it was the beginning of a vast enterprise. His work on radiation effects led him to understand the nature of the tracks: the energetic and highly ionized fission fragments tore through the mica, ripping electrons off the atoms as they went and leaving a record of their passing in the observed tracks, which were then quickly annealed under the bombardment of more electrons in the microscope. But in the twinkling of an eye, another of his interests came to light, one left dangling since his graduate work at Yale involving high-energy nuclear reactions: cosmic rays were also energetic, some of them much more than fission fragments, and might be even more ionizing in the mica. And just down the hall from Walker's lab in Schenectady, two other men had recently shown that chemical etching enlarged atomic dislocations in crystals up to the microscopic level. So, he thought, if mica, or perhaps some other suitable material, were exposed to cosmic rays, tracks would be produced. Chemical etching might preferentially dissolve the mica along the length of the ionized track, both enlarging them so they might be seen in an optical microscope and stabilizing them. The length and diameter of the tracks would be a measure of the energy and chemical composition of the cosmic rays, supplying information of a new and unique kind.

Back in Schenectady, he interested two coworkers, P. Buford Price and Robert Fleischer, both of whom had experience in crystal defect studies, and in the next several years a whole new technology was formed. It would be used to study cosmic rays in meteorites and the lunar surface, fission fragments in deep-sea rocks, plutonium that has long since disappeared from earth, and the effects of nuclear weapons. More to the point, a technique was developed to allow the reading of tracks made by the radioactive decay of heavy elements (such as radium and radon) that emit alpha particles. These, having less energy and being less ionizing, give smaller tracks that are harder to evaluate, and a few false trails were followed, but work in this line continues. An attempt was made to date historical objects, and high-uranium, manmade glasses were dated that were as little as nineteen years old, but the method is not generally applicable. I was able to measure oceanic sedimentation rates, but sophisticated spectrometric techniques proved more valuable. Work has begun using the tracks as a tool to explore for uranium and petroleum and to predict earthquakes—while in the meantime, radon has been growing in importance, in quite a nasty way.

As Europe struggled out of the Middle Ages and began to explore and trade with the East, a demand arose for more metals: gold and silver for trade, iron for weapons. A mountain range on today's German/Czech border was found to be rich in silver, and was named the Erzgebirge, or Ore Mountains. Further mining in the sixteenth century revealed a host of other metals, and when uranium was discovered in 1789 by Martin Klaproth, the digging was extended to include pitchblende, for uranium was found to be useful as a coloring material. The mines, as today, made the owners rich and, much more than today, killed the miners. Contemporary accounts talk of women marrying seven husbands in a row, as one by one the men died, their lungs eaten away by the *Bergsucht,* mountain sickness, which they thought was caused by evil dwarfs—or perhaps by gases from the metals? In another hundred years, lung cancer was recognized as the agent of death, and by the 1930s radioactivity was suspected as the cause.

Radioactivity, harmful to one's health? The suggestion was not embraced by everyone. Madame Curie had fought all her life for the idea that radium was a *cure* for cancer, not a cause. And of course even today controlled doses of radiation are an accepted cancer treatment. The pertinent word there is "controlled," for it is clear that the energies involved in nuclear radiation are roughly a million times more than the energies holding our atoms and molecules together, so that an alpha particle from radon decay can break a DNA molecule in two or knock electrons off atoms, leaving behind rogue DNA to misdirect cell growth or ionized atoms to disrupt the normal chemical processes of our bodies. In radiation therapy, the radiation is carefully directed at the cancer cells, with the hope that they will be destroyed while the surrounding tissue is injured but capable of restoration, resulting in the sickness associated with such therapy but eventually, hopefully, the cure.

Radon is a particularly nasty radionuclide, both because it is an inert gas and also because its decay products—and *their* decay products—are radioactive, giving multiple insults to the body. Uranium miners would naturally be breathing in the stuff, and lung cancer is the result. Well, the world did not get itself in an uproar as this was revealed. Miners of all sorts know the risks, was the general outlook, and we can't worry about a few premature deaths. The world must go on. But the problem has turned out to be much more widespread.

The isotope we're dealing with is radon-222, which results from the decay of uranium-238 and has a half-life of 3.8 days. And uranium is a radionuclide naturally occurring not only in mines but in the soil and rock over which our homes are built, and so is continually producing radon, which, being an inert gas, continually seeps through the ground and wafts into our homes.[4]

And into our lungs. It has two avenues of attack. We breathe the radon in and we breathe it right out again, but this means there is a steady concentration in our lungs, and when it decays the alpha particles rip through the lung tissue like the Wehrmacht through France. And it isn't finished there, for when the radon emits its alpha particle it becomes polonium-218, which emits another alpha to become lead-214. The lead atom is also radioactive, emitting a beta particle, not as damaging as the alpha but still no slouch, and the lead then becomes bismuth-214, which in turn is beta-active and becomes polonium-214, which fires one last alpha particle before settling down as lead-210. All of the radioactive daughters of radon have shorter half-lives and so are in secular equilibrium; that is, their rate of decay is the same as their production rate, so that one radon atom bombards our lungs with three alpha and two beta bullets.

And that's just the first avenue of attack. In addition, the radon floating around the house will decay to polonium. This atom, settling toward the ground or being wafted around in the breeze, is likely to sit down on a speck of dust which might then be inhaled, and which might then be stuck on the lung tissues (instead of being breathed out again like the pure radon atom). So it sits there in our lungs and decays to lead-214 which decays to . . . etc. etc.

All in all, radon is a bitch.

4. The other naturally occurring isotope of uranium, ^{235}U, produces radon-219, but this has a half-life of only a few seconds. Thorium-232 is the other long-lived radon precursor, producing radon-220, with a half-life of just under a minute. Both of these decay before any gaseous diffusion can take place, and so are not a part of the radon problem.

And the bitchiness is compounded because it is so subtle, blowing unseen in the gentle breezes. It was 1959 before the Public Health Service issued any kind of warning, and that only to miners. The danger was just beginning to be talked about seriously in the 1970s, but no one paid any real attention to the possibility of a danger to the general public until 1980, when a Federal task force was organized to examine radon in residential and office buildings. They found that there was "a potential large population at risk" but said nothing should be done pending further studies.

So nothing was done, until December 1984, when Stan Watras set off the alarms at the Limerick Nuclear Generating Station in Boyertown, Pennsylvania.

At the time, the plant was just getting ready to begin operations; the staff were being trained, and radiation detectors had been set up so that everyone leaving the plant had to pass through one in order to get out, to make sure no radioactivity could accidentally be brought out of the plant, but as yet no nuclear fuel had been brought in. So there was no way the detector could detect any radioactivity. But when Stan walked through the detector on his way out, it went Bing! Bing! Bing!

Well, they laughed at first, but then decided they had to follow regulations. So Stan had to strip down, hand over his clothes for washing, shower thoroughly, and then wait around for his clothes to be dried before he could dress again and go home. What a nuisance. And it was just as big a nuisance when it happened again. And again. And again. Finally he decided to prove to them that the detectors were faulty, so when he arrived at the plant the next morning he went through the front doors and then turned right around and went out again through the detectors and—Bing! Bing! Bing!

See, he said, I'm not picking up any radioactivity in here.

No, they agreed, shaking their heads in puzzlement. You're bringing it in.

Oh.

As soon as he thought about this he began to worry. Something in his home? Where else could it be? Last night he had been forced to wash himself and his clothes down before he could leave the plant; he had gone home and hadn't gone anywhere else before coming in to work this morning and setting off the alarms. He told the health physicists they had better go check out his house.

They said no. Their job was to ensure that *the plant* didn't spread radioactivity around; no one had ever said anything to them about other sources.

That's stupid, Watras said.

That's our job, they said.

So he went to their superior, and to another superior, and to another, and finally he found someone who told the health physicists to get the hell out there. That day three of them showed up, and as soon as they walked in the house their Geiger counters started rattling away. They searched for the source of the problem but found that the whole house was equally hot, with just a slight increase in the basement, but with no obvious spill of radioactivity. They wiped down every surface and took the wipes back with them for gamma ray spectrometry, and finally identified the problem.

Polonium, lead, and bismuth.

Radon daughters.

The contamination on his clothes had come about in the same way as the Brookhaven contamination in my lab. Radon swirling around in his home had decayed to polonium, which had settled on his clothing and further decayed to lead and bismuth. The state Bureau of Radiation Protection told the family to get out right away, and the owners of the plant basically took the house apart and put it back together again, and half a year later the family moved back in.

More importantly, this event focused attention on the source of the problem,[5] which is basically a matter of geology. Boyertown lies on the Reading Prong, a large area straddling Pennsylvania, New York, and New Jersey, consisting of granite with a high concentration of uranium. Combined with this is a fractured granitic structure allowing the radon daughter of uranium to seep upwards, and the result is an elevated radon concentration in the homes built there. The combination of high uranium content and open pathways for gas migration is unusual but not unique to the Reading Prong. North central Florida, where homes are built on phosphate deposits, is at risk, as are areas in Illinois, New Mexico, the Dakotas, California, and Washington. The EPA now runs a site devoted to the problem: www.epa.gov/radon.

Luckily, the solution is not part of the problem. In all but the worst cases (where construction repair involving sealing off the basement is required), simple ventilation does the trick, because radon is an inert gas that can be blown out of the house as easily as it leaks in. All that is usually necessary is to ventilate the house thoroughly, wait a few weeks for the radon daughters to die away, and then keep the house well ventilated. The problem lies in knowing how much radon is there, and this is where the GE group comes in. The sophisticated spectrometry the health physicists used to determine the nature of the problem, the same sort of technique needed for measuring oceanic sediments, is overkill. Expensive overkill, much too expensive to be distributed to the tens of thousands of homes at risk. Track detection is the key. The GE group put together a simple detector based on the ability of certain plastics to store the tracks produced by the alpha decays. As now commercially available (the EPA will provide one for

5. The event itself was important enough to the Watras family. Bone scans showed elevated levels of lead-210 in all of them, particularly in the baby. Very scary, with nothing to be done about that, but so far no health symptoms have been reported, although one estimate is that there is a 50% chance of one member of the family dying of lung cancer.

$15), it consists of a cup screened by a permeable membrane at the open end, with a plastic track detector at the closed end. The membrane acts as a filter that keeps out any radioactive particles except for the radon gas, and the detector records the number of decays. After exposure (a few days or months), the cup is sent back to the provider, who etches and counts the tracks and tells you how much radon is in the house. You can then take the necessary steps.

Unless, of course, you *want* to breathe radon.

Believe it or not, some people do. The Merry Widow Health Mine and Earth Angel Health Mine in Montana, former uranium mines, are happy to take your money, along with that of hundreds of other people every year, to allow you to soak your lungs with their radon-filled air and cure you of arthritis, multiple sclerosis, depression, cataracts, migraines, sinusitis, eczema, asthma, hay fever, psoriasis, allergies, diabetes, "and other health problems," according to their web site and an article in the January 2004 *National Geographic*, which quotes Earth Angel's owner, "Wild Bill" Remior, as saying, "Now I seen a dog go in that mine that couldn't hardly walk, and by about the second day he was chasin' rabbits."[6]

There must be something about the wide open spaces of the west. Washington is where UFOs were first sighted, and New Mexico is where, people swear, aliens came to visit.

Well, kiddo, whatever turns you on.

6. Montana regulations restrict these "treatments" to the EPA maximum dose per year, but workers at the facilities can receive far higher doses (as shown by radon-produced tracks in plastic eyeglass lenses).

TWENTY-ONE

L'envoi

The farther backward you can look, the farther forward you can see.
—Winston Churchill

CHURCHILL'S EXPRESSION WAS GLORIOUS RODOMONTADE, but in the end it is still nothing but rodomontade. Understanding the causes of the First World War did not help us to understand the different factors that were operating in 1939, and understanding the results of our isolationism when Hitler began strutting around did not help us avoid the opposite mistakes we made by waging "preventive" war in Vietnam and Iraq. "The past is a different country; they do things differently there," and we learn nothing from them except that we cannot predict the future.

This is true even more with science than with politics. At the end of every century, there is a spate of experts predicting what the new century will bring. But in 1900 no one predicted radio, much less television, or antibiotics or computers or MRI or CAT scans, or cyclotrons or trips to the moon, or even that man might fly. So I cannot pretend that the history written here will tell us what breakthroughs are in store for those working with the noble gases. That's why they call it research; if you knew what the result of your experiment was going to be, there'd be no point in doing it. I thought I knew what the result of Ray Davis's neutrino experiment was going to be, and so I thought there was no point in doing it. I was wrong, and glad to be, for it's the surprises that drive us forward: Rutherford's helium particles bouncing backwards, the xenon-129 peak poking up beyond where it ought to be, the argon-39 peak appearing where it oughtn't to be at all, the electrical currents suddenly running wild through the helium-cooled mercury, et cetera and so forth and so on.

What's coming next? I have no idea and, no matter what they tell you, neither does anyone else. Which is what makes it all so exciting.

Exactly fifty years after I first met the noble gases at Brookhaven in the summer of 1958, I turned off the mass spectrometer and retired. Fifty years to the other side of that summer, a few helium particles had just come bouncing back at Rutherford, in the Netherlands Heike Onnes was beginning his quest to liquefy helium, and the *Wall Street Journal* was convinced there was intelligent life on Mars. Time sure flies when you're having fun.

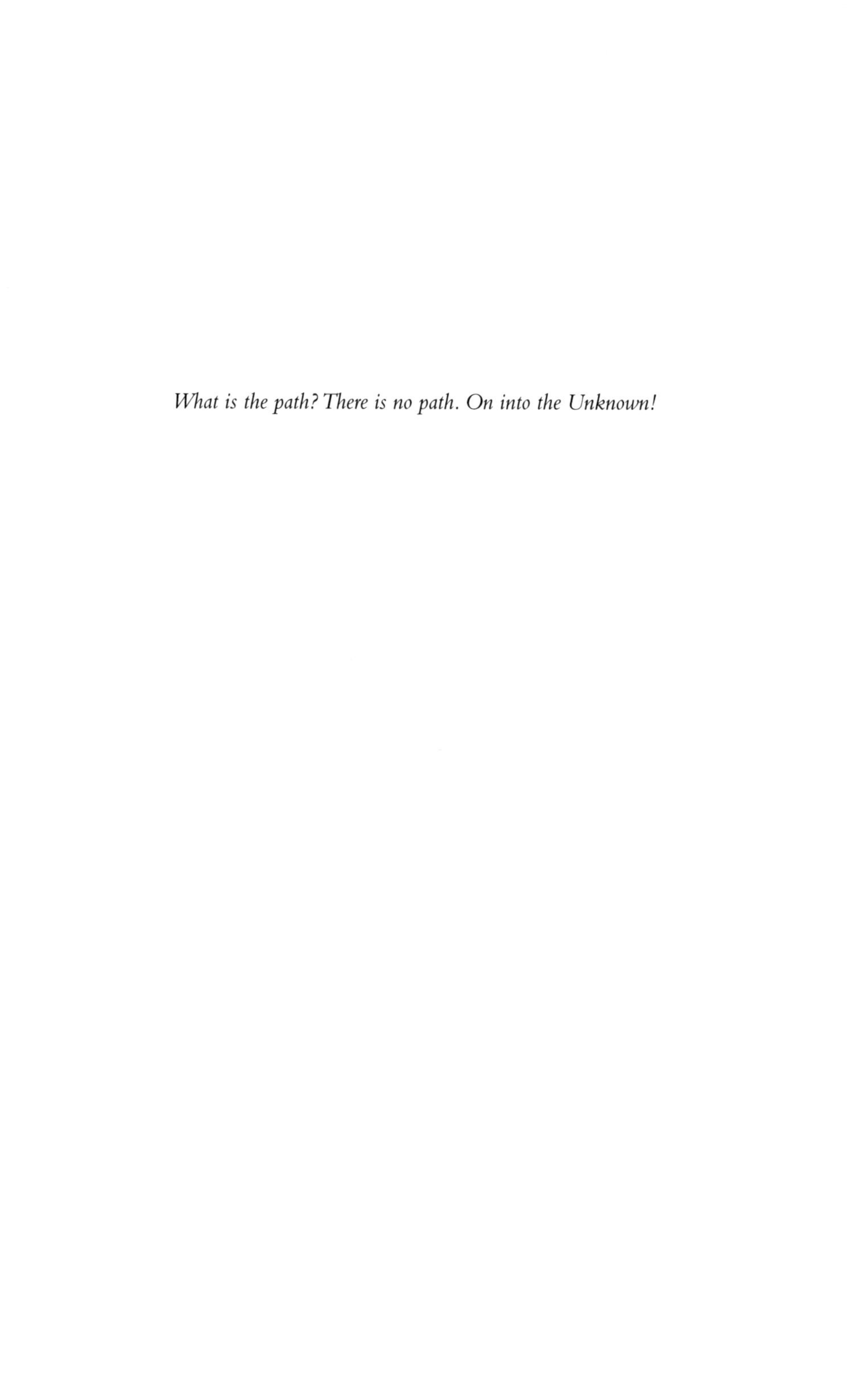

What is the path? There is no path. On into the Unknown!

Notes

Chapter One: Apology

3. "recent advances in the field of geo/cosmochemistry...": D. C. Porcelli, J. Ballentine, and R. Wieler, eds., *Noble Gases in Geochemistry and Cosmochemistry* (Washington: Mineralogical Society of America, 2002). For more background see M. Ozima and F. Podosek, *Noble Gas Geochemistry* (Cambridge University Press, 2002).
3. "plus several chapters in another...": H. D. Holland and K. K. Turekian, *Treatise on Geochemistry* (Elsevier, 2007).
3. "discovery and uses of liquid helium...": D. Van Delft, *Freezing Physics* (Amsterdam: Koninklijke Nederlandse Akademie van Wetenschappen, 2007).
3. "plus another (500 pages...": G. E. Volovik, *The Universe in a Helium Droplet* (Oxford University Press, 2003).
3. "you might want to spend 135 dollars...": Ian McDougall and T. Mark Harrison, *Geochronology and Thermochronology by the $^{40}Ar/^{39}Ar$ Method* (Oxford University Press, 2008).

Chapter Three: Helium

9. "Pierre Jules César Janssen...": http://en.wikipedia.org/wiki/Pierre_Janssen
11. "The air was already being used in warfare...": "Military Use of Balloons in the Mid- and Late Nineteenth Century," available at http://www.centennialofflight.gov/essay/Lighter_than_air/military_balloons_in_Europe/LTA4.htm

12. "helium enemas...": www.dearauntnettie.com/museum/museum-helium.htm

Chapter Four: Argon

18. "The discovery of argon...": Lord Rayleigh, "Argon," *Proceedings of the Royal Institution* 14 (1895): 524–538. Reprinted in Lindsay, *Lord Rayleigh.*
22. "John William Strutt, the third Baron Rayleigh...": J. J. O'Connor and E. F. Robertson, "John William Strutt (Lord Rayleigh)," available at http://www-history.mcs.st-andrews.ac.uk/Biographies/Rayleigh.html; Robert Bruce Lindsay, *Lord Rayleigh: The Man and His Work* (New York: Pergamon Press, 1970).
26. "always be sure to get everything possible...": André Maurois, *The Life of Sir Alexander Fleming* (New York: Dutton, 1959), p. 125.
27. "The scepter passes now from Rayleigh to Ramsay...": http://search.ucl.ac.uk/search?client=default_frontend&site=default_collection&output=xml_no_dtd&proxystylesheet=default_frontend&ie=UTF-8&oe=UTF-8&as_q=+&num=10&btnG=Search+UCL&as_epq=William+Ramsay&as_oq=&as_eq=&lr=&as_ft=i&as_filetype=&as_occt=any&as_dt=i&as_sitesearch=&sort=&access=p&as_lq=
28. "the periodic law...": Carmen J. Giunta, "Argon and the Periodic System: The Piece That Would Not Fit," *Foundations of Chemistry* 3 (2001): 105–128.
30. "An Undiscovered Gas": *Nature* 56 (1897): 378. Address to the Section B (Chemistry) of British Association for the Advancement of Science; reprinted in David Knight, ed., *Classic Papers in Chemistry*, second series (New York: American Elsevier, 1970).

Chapter Five: Helium and the Age of the Earth

33. "Until nearly the end of the nineteenth century...": Two excellent books concerning the age of the earth are G. Brent Dalrymple, *The Age of the Earth* (Stanford University Press, 1991) for pure science and Martin Gorst, *Aeons* (London: Fourth Estate, 2001) for a more popular account.
34. "the entire universe, including our world, was created...": John Lightfoot, *A Few and New Observations on the Book of Genesis* (London, 1642), reprinted in *The Whole Works of the Rev. John Lightfoot D.D.*, ed. John Rogers Pitman, 13 vols. (London: J. F. Dove, 1825).
35. "I judge it indeed difficult but not impossible...": Gorst, *Aeons,* p. 15.
37. "the sacred books of the Hindoos...": *The Autobiography of Charles Darwin: 1809–1882*, ed. Nora Barlow (London: Collins, 1958).
37. "George-Louis Leclerc...": Dalrymple, *Age of the Earth,* pp. 29ff.

39. "which he based on the hydrologic cycle...": For the historical origins of the cycle, see David Oldroyd, *Earth Cycles: A Historical Perspective* (Westport, Conn., and London: Greenwood Press, 2006).
41. "a boy named Ernest now enters upon the scene...": David Wilson, *Rutherford: Simple Genius* (Cambridge, Mass.: MIT Press, 1983); Edward Neville da Costa Andrade, *Rutherford and the Nature of the Atom* (New York: Doubleday, 1964).
43. "The energy produced by the atom is a very poor kind of thing...": John Campbell, "Rutherford—A Brief Biography," available at http://www.rutherford.org.nz/biography.htm
46. "in 1906 he published values of roughly 500 million years...": Dalrymple, *Age of the Earth,* p. 71.

Chapter Six: Helium and the Nuclear Atom

49. "One day Geiger came to me...": Rutherford lecture, quoted in Andrade, *Rutherford,* p. 111.
51. "It's called serendipity...": Walpole was referring to a book he had read as a boy, *Travels and Adventures of Three Princes of Sarendip* (1722), a translation of a 1719 French translation of *Peregrinaggio di tre figluoli del re di Serendippo,* published by Michele Tramezzino in Venice in 1557, in which the princes have many adventures in which they display their keen wits, but it was Walpole who added the concept of "discoveries of things which they were not in quest of." Thanks to copyeditor Steve Dodson for unearthing this information.
52. "as Geiger put it in the paper published in 1910...": H. Geiger, "The Scattering of the α-Particles by Matter," *Proceedings of the Royal Society* A 83 (1910): 492–504.
53. "Supposing that the forces involved...": Edward Neville da Costa Andrade, *Rutherford and the Nature of the Atom* (New York: Doubleday, 1964).
54. "There is a story about him...": The same story is told about the mathematician G. H. Hardy, and probably others as well.
58. "'Shut up,' he explained...": Ring Lardner, "The Young Immigrants," *Saturday Evening Post,* Jan. 31, 1920.

Chapter Eight: Meanwhile, Back at Brookhaven

66. "One mineral—malacone—..." W. Ramsay, "An Undiscovered Gas," *Nature* 56 (1897): 380.
67. "Strongly impressed with the idea that these stones...": Yale Peabody Museum, "The Weston Meteorite," available at http://www.yale.edu/peabody/collections/met/met_weston.html

69. "the uranium-helium ages of several irons...": W. J. Arrol, R. B. Jacobi, and F. A. Paneth, "Meteorites and the Age of the Solar System," *Nature* 149 (1942): 235–238.

69. "the uranium/helium ages were invalid...": Carl August Bauer, "The Absorption of Cosmic Radiation in Meteorites," *Physical Review* 74 (1948): 225–226, and "Rate of Production of Helium in Meteorites by Cosmic Radiation," *Physical Review* 74 (1948): 501–503; H. E. Huntley, "Production of Helium by Cosmic Rays," *Nature* 161 (1948): 356.

69. "Then Fred Singer...": S. F. Singer, "Meteorites and Cosmic Rays," *Nature* (1952): 728.

70. "the origin of the meteorites...": Harold C. Urey, "Origin and Age of Meteorites," *Nature* (1955): 321–323.

70. "So, Nat'ralists observe...": Jonathan Swift, "On Poetry: A Rhapsody" (1733), lines 337–340, *The Works of Jonathan Swift, D.D., Dean of St. Patrick's, Dublin*, vol. 7 (London, 1754), p. 268.

74. "The first attempts...": F. Begemann, J. Geiss, and D. C. Hess, "Radiation Age of a Meteorite from Cosmic-Ray-Produced He^3 and H^3," *Physical Review* 107 (1957): 540–542; E. L. Fireman and D. Schwarzer, "Measurement of Li^6, He^3, and H^3 in Meteorites and Its Relation to Cosmic Radiation," *Geochimica et Cosmochimica Acta* 11 (1957): 252–262.

74. "they could be fragmented asteroids...": D. E. Fisher and O. A. Schaeffer, "Cosmogenic Nuclear Reactions in Iron Meteorites," *Geochimica et Cosmochimica Acta* (1960): 5–14.

75. "space erosion...": F. L. Fireman and E. L. Whipple, "Calculation of Erosion in Space from the Cosmic-Ray Exposure Ages of Meteorites," *Nature* 183 (1959): 1315; D. E. Fisher, "The Origin of Meteorites: Space Erosion and Cosmic Radiation Ages," *Journal of Geophysical Research* 71 (1966): 3251.

Chapter Nine: Cornell, the Ten-Minute Experiment

83. "measuring uranium in iron meteorites...": G. W. Reed, H. Hamaguchi, and A. Turkevich, "The Uranium Contents of Iron Meteorites," *Geochimica et Cosmochimica Acta* 13 (1958): 248.

86. "the uranium content of Sikhote-Alin...": E. L. Fireman and D. E. Fisher, "Uranium in the Sikhote-Alin Meteorite and Its Relation to the Lead Method of Age Determination," *Nature* 192 (1961): 644–645.

88. "ages of ten billion years...": R. W. Stoenner and J. Zähringer, *Geochimica et Cosmochimica Acta* 15 (1958): 40.

Chapter Ten: K/Ar and the Irons

90. "The world is a complicated place...": Bill Watterson, *The Indispensable Calvin And Hobbes* (Riverside, N.J.: Andrews McMeel, 1992), p. 151.

98. "His name now leapt off the pages . . .": G. J. Wasserburg, D. S. Burnett, and C. Frondell, Strontium/Rubidium Age of an Iron Meteorite," *Science* 150 (1965): 3705.
98. "a young age in one iron . . .": David E. Fisher, " 'Ages' of the Sikhote Alin Iron Meteorite," *Science* 139 (1963): 752.

Chapter Eleven: The Spreading Oceans

109 note. "At a conference at Brown University . . .": At the conference on "The Present Interglacial: How and When Will It End?" held at Brown in 1972, J. Murray Mitchell, of NOAA, argued that the emission of CO_2 meant that "the real impact of human activities on the climate of future decades is quite likely to be one of warming."

Chapter Twelve: Dating the Spreading Sea Floor

114. "It was finally translated . . .": A. Wegener, *The Origin of Continents and Oceans* (New York: Dover: 1929, repr. 1966).
116. "Fred Vine . . .": F. J. Vine and D. H. Matthews, "Magnetic Anomalies over Oceanic Ridges," *Nature* 199 (1963): 947–949.
120. "a true Renaissance scientist . . .": Harmon Craig, "Cesare Emiliani," available at http://www.readersadvice.com/mmeade/house/emiliani.html
120. "even the Pope is confused . . .": I. Elliott and C. Emiliani, "Vatican Confusion," *Nature* 375 (1995): 530.
121. "Cesare's Reply . . .": *Journal of Geology* 71 (1963): 810.

Chapter Thirteen: The Argon Surprise

128. "trustworthy K/Ar ages . . .": We tried: D. Fisher, E. Bonatti, O. Joensuu, and J. Funkhouser, "Ages of Pacific Deep-Sea Basalts, and Spreading of the Sea Floor," *Science* 160 (1968): 1106–1107. But further data showed that such ages are not to be trusted due to the possibilities of either excess argon or diffusion loss.
131. "the problem of excess argon in the oceanic rocks . . .": J. G. Funkhouser, D. E. Fisher, and E. Bonatti, "Excess Argon in Deep-Sea Rocks," *Earth and Planetary Science Letters* 5 (1968): 95–100. The discovery was also made independently by two other groups: G. Dalrymple and J. Moore, "Argon-40: Excess in Submarine Pillow Basalts from Kilauea Volcano, Hawaii," *Science* 161 (1968): 1132; C. Noble and J. Naughton, "Deep-Ocean Inert Gas Content and Uncertainties in Age Dating," *Science* 162 (1968): 2675.
132. "Figure 3a shows . . .": H. Brown, "Rare Gases and the Formation of the Earth's Atmosphere," in *The Atmospheres of the Earth and Planets,* ed. G. P. Kuiper (Chicago: University of Chicago Press, 1949), pp. 258–266; H. E.

Suess, "Die Häufigkeit der Edelgase auf der Erde und im Kosmos [The Abundance of Rare Gases in the Earth and in the Cosmos]," *Journal of Geology* 57 (1949): 600–607 (in German).

Chapter Fourteen: Primordial Helium and Argon and the Evolution of the Earth

138. "the (vanishingly small) amount of ^{3}He...": H. Johnson and W. I. Axford, "Production and Loss of He^{3} in the Earth's Atmosphere," *Journal of Geophysical Research – Space Physics* 74 (1969): 2433–2438.

138. "the ^{3}He was primordial...": W. B. Clarke, M. A. Beg, and H. Craig, "Excess ^{3}He in the Sea: Evidence for Terrestrial Primordial Helium," *Earth and Planetary Science Letters* 6 (1969): 213–220.

138. "I published this result...": D. E. Fisher, "Search for ^{3}He in Deep-Sea Basalts," *Earth and Planetary Science Letters* 8 (1970): 77.

139. "Minoru Ozima...": M. Ozima, "Was the Evolution of the Atmosphere Continuous or Catastrophic?" *Nature: Physical Science* 246 (1973): 41–42; M. Ozima and K. Kudo, "Excess Argon in Submarine Basalts and an Earth-Atmosphere Evolution Model," *Nature: Physical Science* 239 (1972): 23–24.

139. "In 1975 Harmon Craig...": J. E. Lupton and H. Craig, "Excess ^{3}He in Oceanic Basalts: Evidence for Terrestrial Primordial Helium," *Earth and Planetary Science Letters* 26 (1975):133–139.

139. "the ^{3}He/^{4}He ratio was quite constant...": See D. Porcelli and K. K. Turekian, "The History of Planetary Degassing as Recorded by Noble Gases," *Treatise on Geochemistry*, vol. 4, ed. Ralph F. Keeling (Amsterdam: Elsevier, 2003), pp. 281–318, and references therein.

140. "Japanese group used a new method..." Nobuo Takaoka and Keisuke Nagao, *Rare-Gas Studies of Cretaceous Deep-Sea Basalts*, Initial Reports of the Deep Sea Drilling Project, Volume LI, LII, and LIII, 135 (1980).

140. "Minoru Ozima and I quickly tried to interpret...": D. E. Fisher, "Trapped Helium and Argon and the Formation of the Atmosphere by Degassing," *Nature* 256 (1975): 113; D. E. Fisher, "Rare Gas Clues to the Origin of the Terrestrial Atmosphere," in *The Early History of the Earth*, ed. B. F. Windley (Hoboken, N.J.: John Wiley: 1975); M. Ozima, "Ar Isotopes and Earth-Atmosphere Evolution Models," *Geochimica et Cosmochimica Acta* 39 (1975): 1127–1134.

140. "a joint U.S.-Japanese conference...": See, for example, Y. Hamano and M. Ozima, "Earth-Atmosphere Evolution Model Based on Ar Isotopic Data," in *Terrestrial Rare Gases,* ed. M. Ozima (Tokyo: Japan Scientific Societies Press, 1978), pp. 155–171; D. E. Fisher, "Terrestrial Potassium and Argon Abundances as Limits to Models of Atmospheric Evolution," ibid., pp. 173–183.

141. "But if we suppose...": F. Albarede "Time-dependent Models of U–Th–He and K–Ar Evolution and the Layering of Mantle Convection," *Chemical Geology* 145 (1999): 413–429; G. F. Davies, "Geophysically Constrained Mantle Mass Flows and the ^{40}Ar Budget: A Degassed Lower Mantle?" *Earth and Planetary Science Letters* 166 (1999): 149–162.

141. "due merely to atmospheric contamination...": D. E. Fisher, "Noble Gases from Oceanic Island Basalts Do Not Require an Undepleted Mantle Source," *Nature* 316 (1985): 716–718.

141. "a separate mantle region...": S. Mukhopadhyay, "Mantle Reservoirs From a Noble Gas Perspective," American Geophysical Union fall meeting 2007, abstract.

141. "comparing the ^{4}He to the ^{40}Ar...": Fisher, "Trapped Helium and Argon."

141. "Ozima argued that the high ^{40}Ar/^{36}Ar...": Hamano and Ozima, "Earth-Atmosphere Evolution Model"; Fisher, "Terrestrial Potassium and Argon Abundances."

142. "Ozima's argument was convincing...": Porcelli and Turekian, "The History of Planetary Degassing."

Chapter Fifteen: Xenology

146. "Wasserburg decided to search for evidence of ^{129}I decay...": G. J. Wasserburg and R. Hayden, "Time Interval between Nucleogenesis and the Formation of Meteorites," *Nature* 176 (1955): 130–131.

148. "a large excess at mass 129...": J. H. Reynolds, "Determination of the Age of the Elements," *Physical Review Letters* 4 (1960): 8–10.

150. "A host of experimental laboratories...": See references in, for a start, F. A. Podosek and T. D. Swindle, "Extinct Radioactivities," in *Meteorites and the Early Solar System,* ed. J. Kerridge and M. Matthews (Tucson: University of Arizona Press, 1988). For more recent results, see M. Chaussidon and M. Gounelle, "Short-lived Radioactive Nuclides in Meteorites and Early Solar System Processes," *Comptes Rendus Geosciences* 339 (2007): 872–884. 163. "more than one process seems to be necessary...": G. R. Huss, B. S. Meyer, G. S. J. N. Goswami, and S. Sahijpal, "Stellar Sources of the Short-Lived Radionuclides in the Early Solar System," *Geochimica et Cosmochimica Acta* 73 (2009): 4922–4945.

150. "an entire conference was held recently...": The papers are gathered in *Geochimica et Cosmochimica Acta* 73 (2009); see the introduction, A. N. Krot and M. Bizzarro, "Chronology of Meteorites and the Early Solar System," pp. 4919–4921.

150. "By applying theoretical constraints...": J. D. Gilmour and G. Turner, "Constraints on Nucleosynthesis from Xenon Isotopes in Presolar Material," *Astrophysical Journal* 657 (2007): 600–608.

150. "nucleosynthesis in a massive star dying...": Chaussidon and Gounelle.

151. "Due to the short half-life of extinct ^{129}I...": T. Staudacher and C. J. Allègre, "Terrestrial Xenology," *Earth and Planetary Science Letters* 60 (1982): 389; C. J. Allègre, J. T. Staudacher, P. Sarda, and M. Kurz, "Constraints on Evolution of Earth's Mantle from Rare Gas Systematics," *Nature* 303 (1983): 762–766.

152. "a range of 75% to 25% early degassing...": D. E. Fisher, "Radiogenic Rare Gases and the Evolutionary History of the Depleted Mantle," *Journal of Geophysical Research* 90 (1985): 1801–1807.

152. "As a later review stated...": Porcelli and Turekian, p. 35.

152. "The variations observed in ^{129}Xe/^{130}Xe...": M. W. Caffee, G. B. Hudson, C. Velsko, G. R. Huss, E. C. Alexander Jr., and A. R. Chivas, "Primordial Noble Gases from Earth's Mantle: Identification of a Primitive Volatile Component," *Science* 285 (1999): 2115–2118.

152. "Reynolds discovered a way to do this...": P. M. Jeffery and J. H. Reynolds, "Origin of Excess Xe^{129} in Stone Meteorites," *Journal of Geophysical Research* 66 (1961): 3582–3583.

152. "Another bit of serendipity at the Reynolds lab...": Turner, private communication, 2009.

152. "following an earlier suggestion...": R. A. Fish and G. G. Goles, "Ambient Xenon: A Key to the History of Meteorites," *Nature* 196 (1962): 27.

153. "a particular meteorite formed...": R. H. Brazzle, O. V. Pravdivtseva, A. P. Meshik, and C. M. Hohenberg, "Verification and Interpretation of the I–Xe Chronometer," *Geochimica et Cosmochimica Acta* 63 (1999): 739–760.

154. "This had actually been noted...": H. Wänke and H. König, "Eine neue Methode zur Kalium-Argon-Altersbestimmung und ihre Anwendung auf Steinmeteorite," *Zeitschrift für Naturforschung* A 14 (1959): 860–866, and a report by T. Sigurgeirsson from the University of Iceland.

155. "K-Ar ages for a large number of meteorites...": T. Kirsten, D. Krankowsky, and J. Zähringer, "Edelgas- und Kalium-Bestimmungen an einer grösseren Zahl von Steinmeteoriten," *Geochimica et Cosmochimica Acta* 27 (1963): 13–42.

156. "These studies are covered in full...": Ian McDougall and T. Mark Harrison, *Geochronology and Thermochronology by the ^{40}Ar/^{39}Ar Method* (Oxford University Press, 2008).

157. "Allègre was free to write a 'Comment'...": T. Staudacher and P. Sarda, "Comment on 'Radiogenic Rare Gases and the Evolutionary History of the Depleted Mantle' by David Fisher," *Journal of Geophysical Research* 92 (1987): 2808; D. E. Fisher, "Reply," *Journal of Geophysical Research* 92 (1987): 2813.

157. "For many researchers and teachers...": Michael Balter, "Allègre Loses Job, Research Split Off," *Science* 287 (2000): 2387.

157. "the ecology of helpless protesting...": Claude Allègre, "The Snows of Mount Kilimanjaro," *L'Express*, Sept. 21, 2006.

Chapter Sixteen: The Coldest Place on Earth

160. "food preservation and...": R. G. Scurlock, *Introduction to History and Origins of Cryogenics* (Oxford: Clarendon Press, 1992).
162. "Kamerlingh Onnes...": See the masterly study by Dirk Van Delft, *Freezing Physics* (Amsterdam: Koninklijke Nederlandse Akademie van Wetenschappen, 2007), pp. 212ff, and the article based on it, "Little Cup of Helium, Big Science," *Physics Today*, March 2008.
163. "a Polish student, Franciszek Groer...": S. Chomet, *New York Times*, Feb. 8, 2000.
175. "superfluidity...": C. A. R. Sá de Melo, "When Fermions Become Bosons: Pairing in Ultracold Gases," *Physics Today*, October 2008.
176. "Cosmic inflation...": G. Brumfiel, "Experimental Cosmology: Cosmos in a Bottle," Nature 451 (2008): 236–238.
177. "the theory which was proposed...": ibid, and J. Bardeen, L. N. Cooper, and J. R. Schrieffer, "Microscopic Theory of Superconductivity," *Physical Review* 106 (1957): 162.
177. "a transition from one metastable energy...": http://en.wikipedia.org/wiki/Neutron_star, and M. A. Alpar, "Pulsars, Glitches and Superfluids," *Physics World* January 1998, available at http://physicsworld.com/cws/article/print/1756.

Chapter Seventeen: Back to the Stars

178. "Fritz Houtermans...": Iosif B. Khriplovich, "The Eventful Life of Fritz Houtermans," *Physics Today,* July 1992, pp. 29ff.
184. "perhaps two helium nuclei could fuse...": William Fowler, interviews by John Greenberg and Carol Bugé, Pasadena, California, May 3, 1983 to May 31, 1984 and October 3, 1986, Oral History Project, California Institute of Technology Archives, available at http://resolver.caltech.edu/CaltechOH:OH_Fowler_W (accessed December 22, 2008).

184 note. "Helium-8 has recently been created...": P. Mueller, I. A. Sulai, A. C. C. Villari, J. A. Alcantara-Nunez, R. Alves-Conde, K. Bailey, G. W. F. Drake, et al., "Nuclear Charge Radius of ^{8}He," *Physical Review Letters* 99 (2007): 252501.

186. "the heavier elements are created...": E. Margaret Burbidge, G. R. Burbidge, William A. Fowler, and F. Hoyle, "Synthesis of the Elements in Stars," *Reviews of Modern Physics* 29 (1957): 547–650 (known colloquially as B^2FH).

188. "Professor Don Brownlee...": D. E. Brownlee, "Cosmic Dust: Collection and Research," *Annual Review of Earth and Planetary Sciences* 13 (1985): 147–173.

188. "an unmanned spacecraft named Stardust...": B. Marty, R. L. Palma, R. O. Pepin, L. Zimmermann, D. J. Schlutter, P. G. Burnard, A. J. Westphal, et al., "Helium and Neon Abundances and Compositions in Cometary Matter," *Science* 319 (2008): 75–78.

189. "one of the few assumptions...": H. E. Suess, "Chemical Evidence Bearing on the Origin of the Solar System," *Annual Review of Astronomy and Astrophysics* 3 (1965): 217–234.

189. "uniform isotopic compositions...": E. Zinner, "Stardust in the Laboratory," *Science* 271 (1996): 41.

189. "well-mixed primordial nebula...": Suess, "Chemical Evidence."

189. "The first anomaly...": J. H. Reynolds and G. Turner, "Rare Gases in the Chondrite Renazzo," *Journal of Geophysical Research* 69 (1964): 3263.

189. "The effects were small...": R. S. Lewis, B. Srinivasan, and E. Anders, "Host Phase of a Strange Xenon Component in Allende," *Science* 190 (1975): 1251.

189. "perhaps mass fractionation...": P. K. Kuroda and O. K. Manuel, "Mass Fractionation and Isotopic Anomalies in Neon and Xenonm," *Nature* 227 (1970): 1113–1116.

189. "fission of an extinct nuclide...": P. K. Kuroda, "Plutonium-244 in the Early Solar System," *Nature* 221 (1969): 726; M. W. Rowe and P. K. Kuroda, "Fissiogenic Xenon from the Pasamonte Meteorite," *Journal of Geophysical Research* 70 (1965): 709–714.

189. "an undiscovered superheavy element...": B. Srinivasan, E. C. Alexander, Jr., O. K. Manuel, and D. E. Troutner, "Xenon and Krypton from the Spontaneous Fission of Californium-252," *Physical Review* 179 (1969): 1166; E. Anders and D. Heymann, "Elements 112 to 119: Were They Present in Meteorites?" *Science* 164 (1969): 821; M. Dakowski, "The Possibility of Extinct Superheavy Elements Occurring in Meteorites," *Earth and Planetary Science Letters* 6 (1969): 152.

189. "perhaps nuclear reactions...": W. A. Fowler, J. L. Greenstein, and F. Hoyle, "Nucleosynthesis during the Early History of the Solar System," *Geophysical Journal of the Royal Astronomical Society* 6 (1962): 148–220; D. E. Fisher, "On the Origin of Fissiogenic Xenon in Meteorites, *Journal of Geophysical Research* 72 (1967): 765.

190. "Perhaps a late supernova...": D. D. Clayton, "Extinct Radioactivities: Trapped Residuals of Presolar Grains," *Astrophysical Journal* 199 (1975): 765; W. M. Howard, M. Arnold, and J. W. Truran, "On Possible Short-lived Progenitors of Fission Xenon in Carbonaceous Chondrites," *Astrophysics and Space Science* 36 (1975): LI.

190. "he had reduced the meteorite...": Lewis, Srinivasan, and Anders, "Host Phase of a Strange Xenon Component."

191. "it was diamond...": R. S. Lewis, T. Ming, J. F. Wacker, E. Anders, and E. Steele, "Interstellar Diamonds in Meteorites," *Nature* 326 (1987): 160.

191. "several other mineral grains...": S. Messenger, L. P. Keller, F. J. Stadermann, R. M. Walker, and E. Zinner, "Samples of Stars beyond the Solar System: Silicate Grains in Interplanetary Dust," *Science* 300 (2003): 105–108.

191. "It took another dozen years...": H. H. Loosli and H. Oeschger, "^{37}Ar and ^{81}Kr in the Atmosphere," *Earth and Planetary Science Letters* 7 (1969): 67–71; Z.-T. Lu, "Atom Trap, Krypton-81, and Saharan Water," *Nuclear Physics News* 18 (2008): 36–40.

191. "a sudden flood of cosmogenic noble gases...": S. Niedermann, "Cosmic-Ray-Produced Noble Gases in Terrestrial Rocks: Dating Tools for Surface Processes," *Reviews in Mineralogy and Geochemistry* 47 (2002): 731–784; H. Craig and R. J. Poreda, "Cosmogenic ^{3}He in terrestrial rocks: The summit lavas of Maui," *Proceedings of the National Academy of Sciences of the United States of America* 83 (1986): 1970–1974; M. D. Kurz, "Cosmogenic Helium in a Terrestrial Igneous Rock," *Nature* 320 (1986): 435–439; D. Lal, "In Situ–Produced Cosmogenic Isotopes in Terrestrial Rocks," *Annual Review of Earth and Planetary Sciences* 16 (1988): 355–388; P. Bierman, "Cosmogenic Glacial Dating, 20 Years and Counting," *Geology* 35 (2007): 575; F. Wang, X. Mao, and C. Liu, "Reconstruction of Climatic Change Quantitatively with Dissolved Noble Gases in Groundwaters," *Chinese Journal of Geochemistry* 25 (2008): 280; K. Ammon, T. J. Dunai, F. M. Stuart, A.-S. Mériaux, and E. Gayer, "Cosmogenic He-3 Exposure Ages and Geochemistry of Basalts from Ascension Island, Atlantic Ocean," *Quaternary Geochronology*, Dec. 9, 2009; A. Srinivasan, Z. Top, P. Schlosser, R. Hohmann, M. Iskenderani, D. Olson, J. E. Lupton, and W. J. Jenkins, "Mantle ^{3}He Distribution and Deep Circulation in the Indian Ocean," *Journal of Geophysical Research (Oceans)* 109 (2004): C06012–6017.

192. "^{26}Al too has been found....": T. L. Ku, L. Wang, S. Luo, and J. R. Southon, "^{26}Al in Seawater and $^{26}Al/^{10}Be$ as Paleo-flux Tracer," *Geophysical Research Letters* 22 (1995): 2163–2166.

Chapter Eighteen: The Neutrino Revolution

195. "Pauli was an otherworldly...": Otto Frisch, *What Little I Remember* (Cambridge, U.K.: Cambridge University Press, 1979).

196. "the elusive neutrino was finally (indirectly) found...": Frederick Reines and Clyde L. Cowan Jun., "The Neutrino," *Nature* 178 (1956): 446–450.

203. "They described the experiment...": J. Bahcall, "Ray Davis: The Scientist and the Man," *Nuclear Physics* B (Proceedings Supplements) 48 (1996): 281–283.

203. "his first result...": Raymond Davis Jr., "Solar Neutrinos. II. Experimental," *Physical Review Letters* 12 (1964): 303.

204. "an upper limit to the sun's...": J. N. Bahcall, "Solar Neutrinos. I. Theoretical," *Physical Review Letters* 12 (1964): 301.

206. "In the early '90s John Bahcall was theorizing...": John N. Bahcall and H. A. Bethe, "A Solution of the Solar-Neutrino Problem," *Physical Review Letters* 65 (1990): 2233–2235; H. A. Bethe and John N. Bahcall, "Solar Neutrinos and the Mikheyev-Smirnov-Wolfenstein Theory," *Physical Review* D 44 (1991): 2962–2969; John N. Bahcall and H. A. Bethe, "Do Solar Neutrino Experiments Imply New Physics?" *Physical Review* D 47 (1993): 1298–1301. (I am indebted to Prof. Neta Bahcall for pointing these out to me.)

206. "many scientists are [still] skeptical of them...": Jay M. Pasachoff, *Astronomy: From the Earth to the Universe* (New York: Harcourt Brace, 5th ed., 1998).

206. "Japan's Super-Kamiokande...": R. Mohapatra, M. Parida, and G. Rajasekaran, "Radiative Magnification of Neutrino Mixings in Split Supersymmetry," *Physical Review D* 72 (2005): 013002.

207. "requires new physics...": John N. Bahcall and Edwin E. Salpeter, "Stellar Energy Generation and Solar Neutrinos," *Physics Today,* Oct. 2005.

207. "This closes experimentally the scientific debate...": Bahcall, "Ray Davis."

207. "he opened the door...": See, for example, F. Halzen and S. R. Klein, "Astronomy and Astrophysics with Neutrinos," *Physics Today* 61 (May 2008): 29–36.

207. "We are even going full circle...": S. T. Dye, "Geoneutrino Measurements and Models Investigate Deep Earth," *Eos, Transactions, American Geophysical Union* 89 (2008): 434–435.

Chapter Nineteen: Life and Death

209. "two scientists at the Navy's Experimental Diving...": A. R. Behnke and O. D. Yarbrough, "Respiratory Resistance, Oil-Water Solubility, and Mental Effects of Argon, compared with Helium and Nitrogen," *American Journal of Physiology* 126 (1939): 409–415.

210. "xenon should be most narcotic...": J. Lawrence, W. Loomis, C. Tobias, and F. Turpin, "Preliminary Observations on the Narcotic Effect of Xenon with a Review of Values for Solubilities of Gases in Water and Oils," *Journal of Physiology* 105 (1946): 197–204.

210. "a pronounced narcotic effect...": Stuart C. Cullen and Erwin G. Gross, "The Anesthetic Properties of Xenon in Animals and Human Beings, with Additional Observations on Krypton," *Science* 113 (1951): 580.

210. "xenon as the anesthetic . . .": The anesthetic mechanism seems to be a selective inhibition of one of the brain's neuroreceptors. B. Preckel, N. Weber, R. Sanders, M. Maze, and W. Schlack, "Molecular Mechanisms Transducing the Anesthetic, Analgesic, and Organ-protective Actions of Xenon," *Anesthesiology* 105 (2006): 187–197.

210. "Inhalation induction is smooth . . .": P. D. Harris and R. Barnes, "The Uses of Helium and Xenon in Current Clinical Practice," *Anesthesia* 63 (2008): 284–293.

211. "In the late 1960s, an imaging test . . .": H. N. Wagner Jr, D. C. Sabiston Jr, J. G. McAfee, J. K. Meyer, and J. K. Langan, "Regional Pulmonary Blood Flow in Man by Radioisotope Scanning," *Journal of the American Medical Association* 187 (1964): 601–603.

213. "Today, the pulmonary arteries can be imaged . . .": B. Amini, C. B. Patel, M. R. Lewin, and R. E. Fisher, "Diagnostic Nuclear Medicine in the ED," *American Journal of Emergency Medicine* (2009), in press.

214. " And so he did . . ." L. Alvarez, *Alvarez*, Basic Books, New York, 1987; J. T. Richelson, *Spying on the Bomb*, Norton, New York. 2006.

216. "the neutron star hypothesized . . .": W. Baade and F. Zwicky, "Remarks on Super-Novae and Cosmic Rays," *Physical Review* 46 (1934): 76–77.

216. "it was subsequently published . . .": T. Gold, "Rotating Neutron Stars as the Origin of the Pulsating Radio Sources," *Nature* 218 (1968): 731.

217. "association of helium with the hydrocarbons . . .": Thomas Gold: *The Deep Hot Biosphere* (New York: Copernicus, 1999).

218. "When the U.S. Bureau of Mines wanted to find . . .": R. D. Munnerlyn and R. D. Miller, "Helium-Bearing Natural Gases of the United States," *Bureau of Mines Bulletin* 617 (1963).

218. "despite the very strong association . . .": W. Dyck and C. E. Dunn, "Helium and Methane Anomalies in Domestic Well Waters in Southwestern Saskatchewan, Canada, and Their Relationship to Other Dissolved Constituents, Oil and Gas Fields, and Tectonic Patterns," *Journal of Geophysical Research* 91 (1986): 12343–12353, and other references in Gold, ibid.

218. "a few Russian geologists . . .": V. F. Nikonov, "O svyazi geliya s neftyanymi uglevodorodami [Relation of helium to petroleum hydrocarbons]," *Doklady Akademii Nauk SSSR* [Proceedings of the USSR Academy of Sciences] 188 (1969): 199–201. Other references in Gold, ibid.

218. "two recent reviews . . .": C.J. Ballentine, R. Burgess, and B. Marty, "Tracing Fluid Origin, Transport and Interaction in the Crust," *Reviews in Mineralogy and Geochemistry* 47 (2002): 539–614; G. P. Glasby, "Abiogenic Origin of Hydrocarbons: An Historical Overview," *Resource Geology* 56 (2006): 85–98.

220 note. "Glycine, an amino acid . . .": Bill Steigerwald, "NASA Researchers Make First Discovery of Life's Building Block in Comet," NASA Goddard Space Flight Center, Aug. 17, 2009, available at http://www.nasa.gov/mission_pages/stardust/news/stardust_amino_acid.html

221. "all of these come from levels much deeper...": T. Gold and M. Held, "Helium-Nitrogen-Methane Systematics in Natural Gases of Texas and Kansas," *Journal of Petroleum Geology* 10 (1987): 415–424.

221. "an independent report...": Dyck and Dunn, "Helium and Methane Anomalies."

221. "In the 26 July 2009 issue of *Nature*...": A. Kolesnikov, V. G. Kutcherov, and A. F. Goncharov, "Methane-Derived Hydrocarbons Produced Under Upper-Mantle Conditions," *Nature Geoscience* 2 (2009): 566–570; V. G. Kutcherov, and V. A. Krayushkin, "Deep-Seated Abiogenic Origin of Petroleum: From Geological Assessment to Physical Theory," *Rev. Geophys.* 48 (2010): RG1001.

223. "The most extraordinary event of the year...": D. E. Fisher and M. J. Fisher, *Strangers in the Night* (Washington D.C.: Counterpoint Press, 1998), pp. 41ff.

223. "On July 20, 1976...": For further details see, for example, Fisher and Fisher, *Strangers in the Night.*

223. "We conclude that they [the data] are evidence...": David S. McKay, Everett K. Gibson Jr., Kathie L. Thomas-Keprta, Hojatollah Vali, Christopher S. Romanek, Simon J. Clemett, Xavier D. F. Chillier, Claude R. Maechling, Richard N. Zare, "Search for Past Life on Mars: Possible Relic Biogenic Activity in Martian Meteorite ALH84001," *Science* ≪issue number?≫ (1996): 924–930.

224. "melted less than a billion years ago...": D. Heymann, E. Mazor, and E. Anders, "Ages of Calcium-Rich Achondrites—I. Eucrites," *Geochimica et Cosmochimica Acta* 32 (1968): 1241–1268.

225. "In 1980, two graduate students...": H. McSween and E. Stolper, "Basaltic meteorites," *Scientific American*, June 1980, pp. 54–62.

225. "A martian derivation...": D. Walker, E. Stolper and J. Hays, "Basaltic Volcanism: The Importance of Planet Size," *Proceedings of the Tenth Lunar and Planetary Science Conference* (Oxford and New York: Pergamon Press, 1979), pp. 1995–2015.

225. "A martian derivation...": L. Nyquist, D. Bogard, J. Wooden, H. Wiesmanjn, C-Y. Shih, B. Bansai and G. McKay, "Early Differentiation, Late Magmatism, and Recent Bombardment on the Shergottite Parent Planet," *Meteoritics* 14 (1979): 502.

226. "the noble gases in a fourth Shergottite...": D. D. Bogard and P. Johnson, "Martian Gases in an Antarctic Meteorite?" *Science* 221 (1983): 651–654.

226. "an excellent fit to the abundances...": R. Pepin, "Evidence of Martian Origins," *Nature* 317 (1985): 473.

227. "Plumes of methane...": M. J. Mumma, G. L. Villanueva, R. E. Novak, T. Hewagama, B. P. Bonev, M. A. DiSanti, A. M. Mandell, and M. D. Smith,

"Strong Release of Methane on Mars in Northern Summer 2003," *Science Express* 323 (2009): 1041, quoted in *Physics Today*, April 2009, p. 18.

228. "a pleasant way to ease your path...": ergo, "More Ways to Get the Helium Hood Kits," Assisted-Suicide Blog, Jan. 2, 2009, available at http://assistedsuicide.org/blog/2009/01/02/update-more-ways-to-get-the-helium-hood-kits/

Chapter Twenty: Radon and You

232. "it's 1954 at the General Electric...": R. L. Fleischer, *Tracks to Innovation* (New York: Springer, 1998); T. J. Bernatowicz, "The Multifarious Scientific Career of Robert M. Walker," *Geochimica et Cosmochimica Acta* 67 (2003): 4661–4664.
234. "to predict earthquakes...": Peter N. Spotts, "Can Radon Gas Leaks Predict Earthquakes?" *Christian Science Monitor,* April 8, 2009; M. Singh, M. Kumar, R. K. Jain, and R. P. Chatrath, "Radon in Ground Water Related to Seismic Events," *Radiation Measurements* 30 (1999): 465–469.
234. "the Erzgebirge...": Robert K. Lewis, "A History of Radon- 1470 to 1984," paper presented at the 2006 National Radon Meeting, available at http://www.crcpd.org/Pubs/Radon/HistoryOfRadon.pdf
237. "when Stan Watras set off the alarms..." : Lewis, "A History of Radon."
240. "The Merry Widow Health Mine...": Kira Salak, "ZipUSA: 59631," *National Geographic* Online Extra, January 2004, available at http://ngm.nationalgeographic.com/ngm/0401/feature7/fulltext.html
240. "Montana regulations restrict...": R. L. Fleischer, private communication.

Chapter Twenty-One: L'Envoi

241. "The past is a different country...": L. P. Hartley, *The Go-Between* (London: Hamish Hamilton, 1953).

Index